G. T. EMERY

A Matter
of Choices

LIVES OF WOMEN IN SCIENCE

Founding Editor: Pnina Abir-Am
Series Editor: Ann Hibner Koblitz

VOLUMES IN THE SERIES

*A Convergence of Lives: Sofia Kovalevskaia—
Scientist, Writer, Revolutionary,*
by Ann Hibner Koblitz

*A Matter of Choices: Memoirs
of a Female Physicist,*
by Fay Ajzenberg-Selove

*"Almost a Man of Genius": Clemence Royer,
Feminism, and Nineteenth-Century Science,*
by Joy Harvey (forthcoming)

A Matter of Choices

Memoirs of a Female Physicist

FAY AJZENBERG-SELOVE

RUTGERS UNIVERSITY PRESS
New Brunswick, New Jersey

Library of Congress Cataloging-in-Publication Data

Ajzenberg-Selove, Fay, 1926–
    A matter of choices : memoirs of a female physicist / Fay
Ajzenberg-Selove.
        p.   cm. — (Lives of women in science)
    Includes bibliographical references and index.
    ISBN 0-8135-2034-7 (cloth) — ISBN 0-8135-2035-5 (pbk.)
    1. Ajzenberg-Selove, Fay, 1926– 2. Nuclear physicists—
United States—Biography.   3. Women in science—United
States.   I. Title.   II. Series.
QC16.A34A3   1993
539.7'092—dc20
[B]   93-28136
                                                                      CIP

British Cataloging-in-Publication information available

Contents

Series Editor's Foreword

In the past ten years or so, theoretical feminists have been paying increasing attention to the interactions of gender, science, and culture. Authors such as Evelyn Fox Keller (*Reflections on Gender and Science*, 1985) and Sandra Harding (*Whose Science? Whose Knowledge?*, 1991) are representative of an influential group of theorists who claim that modern science is inextricably linked to masculinist ideals. For them, being a woman scientist is almost a contradiction in terms; it is possible, according to Keller, "only if she undergoes a radical disidentification from self."

This theorizing has generally been at an extreme level of abstraction, rarely based on concrete historical study of the lives of actual women scientists. One cannot avoid the impression that these gender and science writers are not cognizant of the nature and the extent of women's participation in the history of science. As a result, they are unaware that the lives of women scientists often call into question the basic tenets of gender and science theory.

In part, this failure of historical perception of Keller, Harding, and similar authors is understandable. The written record of scientific achievement has been characterized by a plethora of biographical materials about (white) male European and North American scientists, with very few accounts of the scientific accomplishments of women, or of people from Asia, Africa, and Latin America. Feminist gender and science theorists have, perhaps unconsciously, made the assumption that the scarcity of published material about women scientists is equivalent to a scarcity of female contributions to the scientific enterprise, and have evolved their theories accordingly.

The Rutgers University Press series Lives of Women in Science is an attempt to redress the gender imbalance in the historical record, and make available to a wider public the fascinating and varied life stories of women in the scientific community. The present volume, the autobiography of the distinguished nuclear physicist Fay Ajzenberg-Selove, is an exciting contribution to the series. Ajzenberg-Selove's career illustrates well the complexity of the interactions of gender and science in twentieth-century Western society. She was certainly discriminated against, at times rather crudely, by some of her departmental colleagues (especially at the University of Pennsylvania). Yet, she bears them no grudge, because she and her fellow physicists clearly have a strong feeling of camaraderie coming from shared cultural as well as scientific values.

For Ajzenberg-Selove and her peers, the research milieu in physics is a beautiful, exciting, yet comforting world. It is far from being cold or alienating; far from depriving her of her femininity; far from excluding her from full participation. Ajzenberg-Selove's narrative is permeated with joy and deep satisfaction with the role that physics has played in her life. As she says: "Science is not a dead cathedral. It is live and it is *fun*, and it is full of passion."

ANN HIBNER KOBLITZ
Seattle, Washington
May 1993

Acknowledgments

I am very grateful to the people who encouraged me to write this book, and who gave me advice and criticisms. I am particularly indebted to Marcia Bartusiak; to Alice Calaprice; to the series editor, Ann Hibner Koblitz; and to the science editor of Rutgers University Press, Karen Reeds, as well as to my sister, Yvette Louria, and to my husband, for their remarks. All of them urged me to tone down the robust language which I had used in my original manuscript and, somewhat regretfully, I did.

A Matter
of Choices

Introduction

I AM A PROFESSOR OF PHYSICS at the University of Pennsylvania. I am sixty-seven years old. When I began to work in physics, only one in forty American physicists was a woman; now the number is about one in ten. In the United States, fewer women work in physics than in any other scientific field. Why is this the case? There has always been gender discrimination in all professions, in all countries; but overt discrimination has not been worse in physics than in other fields. So why are there so few female physicists?

I think that the traits that are particularly pronounced among successful physicists (who are of course the role models) are unappealing to young women (and, indeed, to some young men) who might consider physics as a career. Being a physicist is fun, but it is not an easy road—not for a man, not for a woman. The social structure of physics is much like a pyramid, with a few successful people at the top, and many others below. It is hard to make one's mark in the field, whether one is a man or a woman. Physicists think of themselves as part of a super-elite, working at the frontiers of science. They show an obsessive single-mindedness in their work; and they are intensely competitive.

Women are no less gifted than men in the ability to do scientific work, but society discourages women from enjoying competition, and winning, except in sex-segregated sports. Women who do better than men are still likely to be labeled as aggressive, strident, hard to work with, and unfeminine.

But women *have* become good physicists. Why we made it, and why we enjoy being physicists, cannot be explained through statistical tables. Each of our stories is different. Each is equally valid.

This is my story. I started to write this book to try to understand why I came to be what I am, and I gradually began to feel that my experiences might be of interest to others.

I came to the United States as a refugee, during World War II. My father was the dominant person in my early life. He wanted me to be an engineer, as he was, but in college I switched to physics. My academic work was extremely poor but I loved physics, and I was determined to succeed. Several people helped me at crucial junctures in my life and made it possible for me to survive as a physicist.

I became a nuclear physicist and a teacher. I still teach physics and I feel excited and revitalized by my students. My personal life has been extremely happy: I have been married for thirty-seven years to a physicist who is as driven as I have been, and who loves me as I love him.

This is basically a simple story. It is the story of a woman enthralled by a romantic view of physics, one who likes adventure, is very stubborn, and very competitive. It is also the story of a woman who escaped the Holocaust in Europe and survived some serious medical problems later in life. But at its most basic, this is a love story: a love of life, of friendship, of physics, and of my husband, Walter Selove.

1

La Sucrerie de Lieusaint

1926–1940······················

Lieusaint is twenty miles south of Paris, on the road to Fontainebleau. In the 1930s it consisted of a single street with grim, undistinguished houses with dovetailed façades that were broken occasionally by a small store. At the back of the houses were individual vegetable gardens, separated from one another by high fences. One of the main roads to the south, the RN 6, went through Lieusaint. To the east stood a large factory, dark and alien, surrounded by lush, green fields of sugar beets. It was bordered by railroad tracks that emerged from the Gare de Lyon. There was a workers' café and, within a mile, the village of Moissy-Cramayel.

I lived in France from 1930 until 1940. For most of that time my father was the director of this factory, the Sucrerie de Lieusaint, a sugar refinery that processed the sugar beets of the region. We lived for a short time in a house in Lieusaint, and then moved to a new cottage in Moissy-Cramayel. I went to primary school there for a while, and while the cottage was my father's main domicile, we had a series of apartments in Paris, and I went to school primarily there. To the extent that I think of any place in France as home, it is of the Sucrerie and Moissy that I think rather than of Paris. Paris was the place where Mother lived, but I adored my father, and I grew up following in his footsteps, striving to like the things he liked. I walked with him around the factory, enjoying the beauty of the machinery, smelling the sugar pulp, and watching his confident and comfortable discussions with the workers. In Moissy, I built model airplanes, listened to political discussions, climbed onto the roof of a garage with a band of boys, raised a couple of chickens, and planted three fruit trees.

And I dreamed of travel, of becoming a pilot, of visiting the United States, and, above all, of becoming an engineer like my father. I have remained a daydreamer all my life. I dream every day—in my quiet home, during boring lectures, on interminably long plane trips, and even when I should be working efficiently and logically. And most of my dreams have come true—the good ones of science, of friendship, of fun. But so have the nightmares of illness, depression, and loss.

My father was born in Warsaw around 1890, in a very poor family. I don't know what my grandfather did. I was told that he was very pious and that he died after fasting on Yom Kippur. My father had two sisters and four brothers. One of his brothers died young, and one of his sisters was killed in the Holocaust. At the time my father was born, Warsaw belonged to Russia. He was an excellent student and he received a scholarship from the czar to attend the St. Petersburg Mining Academy, then the premier engineering school in Russia. The scholarship was extremely unusual: it was difficult for Jews to enter universities, and even more so for poor Jews without connections. Most Jews lived in shtetls, and few were permitted to live in St. Petersburg. But Father was tough, smart, and adaptable. He spoke Polish and Russian at that time, as he was later to speak German, French, and English, and his name changed to meet his location. In Russia it was Moisei Abramovich Aisenberg; in Germany it was M. A. Eisenberg; in France, Michel Aizenberg; and in the United States, Mike Ajzenberg. We called him Misha, everywhere.

My mother, Olga Naiditch, was born in Pinsk, which

was in Russia at the time. Her parents died young, and she moved to St. Petersburg where her older brother, Isaac, lived. She had four older sisters who became her surrogate mothers. Isaac Naiditch had been permitted to live in St. Petersburg because he had built a very large business that produced alcohol. He was a good businessman, but he was also a wise and subtle thinker, part intellectual and part politician. He was also very domineering and a womanizer. In addition to his wife and to his mistress-en-titre (her nickname was Madame Pompadour), he also had a series of quite open short liaisons with younger women. Isaac was one of the early Zionists and organized the self-taxation scheme to which Jews contributed to buy land in Palestine. Mother adored him, and I think that he cared deeply for her. She was about the age of his own children and, in temperament and intelligence, far more interesting than any of them. In St. Petersburg, Mother completed her secondary education and began studies at the Academy of Music. She was a mezzo-soprano and played the piano, but she had neither the discipline nor the hunger for a career. She was a beautiful, spoiled, and very emotional young woman. My parents met, fell in love, and married very young. My sister, Yvette (Iva), was born when they were twenty-one. Mother spoke of happy times, of the White Nights of spring, and of vacations on the Aland Islands between Finland and Sweden.

Misha was very proud of his degree in mining engineering. I still have his certificate: he received the mark "Excellent" in virtually all courses, and he was awarded a special prize for his academic work: a very heavy bronze statuette of a miner on a marble base. (During World War II, Misha buried it in the garden of our cottage and retrieved it later.) After he received his degree, Misha worked as a miner, as was required by his scholarship. Mother told me that he became gravely ill from dysentery due to the appalling conditions in which the workers lived. His stint in the mine gave him an understanding of the problems of blue-collar workers. He was staunchly anti-union and very conservative politically, but he always treated his workers with respect. Then came World War I and the Russian Revolution, and the end of my family's quiet life.

Misha died suddenly in 1962, and in his last years I had been too preoccupied with the beginning of my married life, my first house, and my first permanent job to ask him for details of what occurred between 1919 and 1930. In 1930 we moved to France, and I began to have memories of my own. Mother talked about that period, but her recollections changed with her moods. She told me that at the time of the Revolution, Misha worked as an engineer in Lugansk (later called Voroshilovgrad, and now Lugansk again), in the eastern Ukraine, and that she lived with him there with Yvette and a nurse, of whom she was very fond. The living and political conditions deteriorated, and Misha hijacked a locomotive at the point of a gun, put her and Yvette on it along with some valuables, and drove it through the Red and the White lines. The nurse was a Communist and stayed. They continued to travel west, and landed in Berlin. Inexplicably, Misha became an investment banker there. Four years later he owned a company and a very luxurious country estate. I have a photo of it, and I know the name of the villa—the "Cupola Villa" in Neubabelsberg. I glance at the picture often because it puzzles me. I cannot fit it into my memories of Misha, who loved beauty but was a modest and quite ascetic man. I assume that he got the villa to please my mother.

Shortly after settling in Germany, Mother left Misha and went to live in Paris with Yvette. Her family, which was headed as always by Isaac, had left Russia and settled there. Mother refused to speak of this period, which may mean that the split was due to a love affair, either on Misha's side or on hers. Silence, in our house, as indeed in most Russian-Jewish homes, usually meant that sexual matters were involved. It is hard for me to understand this now, but the belief that sex was dirty, that it shouldn't be talked about, that it was somehow unpleasant and abnormal, was so strongly inculcated in me, as in other girls at that time, that I did not know how a baby came about, nor anything about sexual organs, male or female, until my twenties! Even my first menses was a shock to me, but I did not question Mother's statement that it was normal, and I didn't inquire further. The curiosity I felt toward most other aspects of life stopped at sexual matters. I have read thousands of books, and I always manage to skip over any mention of sexual acts: my mind censors them. In retrospect, while I am intellectually appalled by my parents' view of sex, I realize that my upbringing has served me well. My sexual inhibitions prevented me from having casual intimate relationships, and this, in turn, helped me to have deep and wonderful platonic friendships with men in my almost entirely male field.

After some months, my parents reconciled and Mother returned to Neubabelsberg, and I was born in 1926 at 47 Mommssenstrasse in Berlin. My sister named me after an English actress, Fay Compton, whom she admired. I remember virtually nothing of my life in Germany, except for some sensual feelings when I was bathed in a step-down bathtub at Neubabelsberg. In photographs, the house is shown filled with oriental rugs, and Mother and her friends are fashionably dressed, looking terribly bored. Misha was a polymath and took pleasure in beautiful objects: he designed the silver settings (and had them cast under his supervision) as well as some of the furniture. He bought Sevres and Rosenthal porcelain and he commissioned a Steinway piano of inlaid wood for Mother. (When he became bankrupt during the German Depression, most of these items were placed under court seals. In the mid-thirties, when Misha was able to repay his debts, many of them were returned to us and, as I grew up, I would find forgotten seals under the furniture.) We left Berlin for Paris in 1930. I have a picture of myself then, with curly dark hair and a very sad expression, which is not at all how I remember this part of my childhood. I was indulged and loved, despite my precocious and often tactless remarks. I did not understand, of course, the problems of the times, the Great Depression, the coming of Nazism, my father's bankruptcy, and the need to move west, to France. For me it was an adventure. I was four years old. My biggest shock was when Yvette told me on the train to France that my two stuffed animals, Maya and Teddy, in whom I had confided everything, could not actually see the interesting countryside.

One constant in my life was the Russian language. While German was spoken in the kindergarten I had attended in Berlin, my parents spoke Russian at home and with most of their friends, and I soon picked up enough of the language to be able to eavesdrop on them.

By the time we arrived in France, Isaac had become quite wealthy again, and once again he dominated our world. Mother's sisters surrounded us with love, Isaac helped us financially and placed Misha in charge of the Sucrerie de Lieusaint. But once again my father was a poor relation. In retrospect, I don't think that this bothered him because I think that he was really a very self-confident man. But I am sure that it bothered Mother a great deal: she had regressed from being the hostess of the Cupola Villa to living in a cottage and a small apartment. She felt she had lost face. And so she did things that were understandable but devastating to the fabric of her marriage. She became a compulsive buyer of unsuitable and expensive clothes and other objects. She insisted on household help, and when Misha told her there was not enough money, she used to scream at him, borrow money, and then repay the loans with monetary gifts from Isaac.

When Mother's troubles became severe and the pyramid of loans was too heavy to deal with, she would threaten to kill herself. She would show me boxes of sleeping pills, which I would beg her not to use and which I would try to hide. Much of my childhood was full of the noises of her screaming, and of her threats of suicide. When Mother screamed at him, Misha listened quietly, and then he would leave the house. I was so worried that he would go away permanently, and I blamed myself because some of her extravagant purchases were for me. She once bought me a very expensive bright red coat with a beaver collar, which was quite unsuitable for me, particularly when every French girl my age wore dark, plain clothes (and I so desperately wanted to be one of them and not stand out). When I said that I did not want to wear it on the street, she replied that if I refused she would start to sing loudly in public. This would have been even more traumatic than wearing the despised coat, and so I gave in to the blackmail. This was a mistake, because from then on she would threaten to sing in public unless I did any number of other trivial things. I can't remember if she actually ever did so.

I realize that in referring to Olga as "Mother" and to my father as "Misha," I have shown the way I feel about them. I realize that I may appear to idealize the one and to denigrate the other, and that many daughters have a more difficult interaction with their mothers. In some cases this is surely because the parent who primarily raises them is the mother, and the father is likely to be less judgmental of a daughter than of a son. I think that this is not a strong factor in my case. Both my parents loved me very much and were interested and involved in raising me; and since I did not have a brother, Misha judged me as if I were his son. Mother did have strong qualities: she was extremely emotional, but one of her emotions was surely love for me. She also knew how to hate: in memory of the pogroms in Russia, she discreetly spit three times (clicking her tongue against her teeth) whenever she saw a priest, be he Catholic or Orthodox. Mother was extremely bright, read a great deal, enjoyed music, and had a magnificent voice. She sang folk songs and ballads and her husky, mezzo voice was mesmerizing. She taught me to be cynical about people, and to be devoted and loyal to those who were good. She taught me, above all, that an intelligent woman could not just live through her family, and that a bored woman could be enormously destructive to herself and to the people around her. My gratitude to her for what she gave me of herself and what she taught me to avoid is real. My sister, whom she often treated shabbily, and I took care of her needs over the years, and when she died, at age ninety, I felt no grief. I felt that I had repaid my debt to her, and I was relieved that I would no longer have to listen to her emotional outbursts.

On a typical day in Paris I went to school through the middle of the afternoon. In the later years we lived at 8, Rue César Franck, and I went to school at the Lycée Victor-Duruy located half a mile away. The apartment was on the third floor of a typical stone-faced apartment building, at the corner of the Rue Bouchot. When you opened the door of the apartment after braving the concierge on the first floor, your first view was of Mother's Steinway in the living room, which had bay windows and overlooked both streets. To the right of the entrance was a T-shaped corridor. On the left was Misha's den, filled with a large desk, bookcases, and the sofa on which he often slept. To the right was a kitchen that overlooked the courtyard and the back stairs. At the apex of the T was the bathroom. My room, which faced on the street, and Mother's room, which faced onto a back court, complete the floor plan in my mind. I remember my room as large and cheerful, with plain modern furniture, painted yellow, which had been designed by Misha in Germany. On the walls I had posters of my heroes (the most important of whom was the pilot Amelia Earhart). I had many books and a splendid bed which was high enough that, even in my early teens, I could crawl underneath it and, with a throw draped over the bed, find myself in a cavern of my own, protected from the rest of a complicated world. That was my favorite haven.

Mother's room had a wide, low bed, and it was usually dark, with the drapes drawn and with a strong smell of the cigarettes she chain-smoked, mixed with her favorite perfume, "Jasmin de Corse." The dominant color of the room was mauve, consistent in my mind with my mother's neurotic moods. It was there that she would tell me that she would soon kill herself.

The Lycée Victor-Duruy on the Boulevard des Invalides

looked then, as it does now, like a prison. A stone building with a high stone wall surrounded the inner buildings and the recreation courtyard. It was an all-girls school, staffed by women, rigorous and dull in its teaching and firm in its discipline. Here I was taught to memorize French history, classical French literature, simple mathematics, and English vocabulary, in addition to spelling and geography, and the place gave me a lasting dislike for pedantic authority figures. When we were not writing, we were required to sit quietly, with our arms folded in front of us. (At an earlier, less liberal school, we had to cross our arms behind our backs.) I do not recall any science courses. We were taught English by memorizing every evening the next one hundred words in the dictionary. We had to know the words' spelling and their meaning, but pronouncing them correctly was not part of the curriculum. *Dictées* in French were a daily part of the curriculum, as was a pervasive jingoism directed against the historic enemy, Germany. But much more vicious were the attacks against England and, for that matter, against all who had not been French for at least a hundred years. It is surely not surprising that I did not have any friends at the lycée, and I was never invited to anyone's home. As a child, I assumed that the reason my classmates were not friendly was my fault, and I did not realize until I came to the United States, as a refugee with penniless parents, that I was not intrinsically a pariah.

But I do not mean to suggest that my life was grim. It was in fact very interesting. My freedom to be on my own was nearly total, thanks to Mother's sister, Sara. Sara was a Freudian psychoanalyst who had a small practice. She was some ten years older than Mother and had taken care of Mother when she orphaned. Sara left Russia as a young woman to complete her medical degree in Zurich and then studied with an assistant to Freud before becoming an analyst. She did not marry. In addition to her work, she took care of another sister, Fania, who was a tiny, somewhat feeble-minded woman. Even that care was not undertaken in a traditional way—Fania lived in a small apartment of her own. Sara was very independent and understood my need for freedom. She persuaded Mother to let me travel on my own through Paris to visit her at any time. I saw her nearly every week. She taught me how to analyze my feelings honestly, and how to explain my wants and needs convincingly to my parents. We loved and respected each other.

And I learned how to use the freedom that I gained through Sara. From the age of nine I traveled alone all through Paris, using my allowance to go by bus or metro to the bookstores of the Boulevard St. Michel, to the Palais de la Découverte (a science museum which, at the time, was a treasure), to the Jardin D'Acclimatation (a park with a zoo and rides), and, in 1937, to the World's Fair, to which I had somehow obtained a season pass. The excitement of the fair was tremendous, with the huge pavilions of Nazi Germany and Stalinist USSR facing each other. At mealtimes I did not hesitate to get a snack at a store, or even at a café. I read many books, primarily the classics, but I was most deeply affected by contemporary works such as *Man's Hope* and *Man's Fate* by André Malraux, which were about the Civil War in Spain and the revolution in China. I was also touched by romantic stories in which the hero (or the heroine) comes to the end of his life and realizes that he had wanted quite a different life than the one he had led. I consciously resolved to live a life that I would not regret as I lay dying.

My allowance was also spent on cigarettes for another woman who had a very strong influence on me. Her name was Lida Nahimovich. She was the cousin of the wife of one of Isaac's sons. Lida was a beautiful young woman, about eight years older than I, a some-time student at the Sorbonne and a Communist. She was penniless and living with a young philosophy student, a Frenchman named Jean-Pierre Vernant who subsequently became a Resistance leader ("Colonel Berthier") and later a well-known professor at the Collège de France. They married during the war. For reasons I will try to explain later, I did not remain in touch with them for well over forty years. In my childhood she gave me many things I needed: a sense of concern for less privileged people, a sense of fun as she recounted her experiences, a feeling for the beauty of the senses; for a while, I totally idolized her. I imitated her handwriting and I tried to look like her, without success. Lucky Strikes cigarettes were her favorite, despite the fact that they were American—she disliked capitalist America as she worked for the socialist Nirvana. Lida was a glamorous and wonderful creature to me: she taught me to like the poetry of Verlaine and of Garcia Lorca, and she taught me the songs of the Auberges de la Jeunesse one of which I still sing to myself, as a talisman, when I am scared: "Unissons nos voix avant de nous quitter. Je vais parcourir d'autres lieux. La vie est si douce et le monde si beau, entonnons ce dernier adieu" (Let us join our voices before separating. I will travel across other lands. Life is so sweet and the world so beautiful, let us sing to this last parting). The last stanza is "Je serai prête pour mon dernier voyage. Je dirai mon dernier adieu" (I will be ready for my ultimate voyage. I will say my final goodbye).[1]

In the context of a twelve-year-old in a reasonably

peaceful era, these words are romantic and meaningless. In the thirties in Europe they were not. Beginning with the reoccupation of the Rhineland in 1936, events moved quickly: the Civil War in Spain, the Anschluss with Austria, the annexation of the Sudetenland, and the foolish refusal of the Western democracies to prevent the continued rise of Hitler, to stop him when it could still be done easily. Isaac and his Zionist friends, who included Chaim Weizmann, were very much aware of the horrors to come. It was completely clear to all of us, even to me, the youngest member of our extended family, that war would occur. From 1936, when I was ten years old, I lived with a sharp awareness that I was unlikely to survive my teens. And there was really no place to escape to at that time.

We were less aware of the dreadful events happening in the USSR under Stalin. Russia was farther away than Germany, its propaganda was more believable than that of the Nazis; it appeared to be fighting against the Nazis in Spain, and it did not appear to single out Jews for oppression and death. We should have known, but we did not, that there was no substantial difference between Hitler and Stalin. The German-Soviet pact in August 1939, which led to the dismemberment of Poland and which signaled the formal start of World War II, was a cruel revelation.

Still, France was a beautiful country. There were fun things to do, and one can live with the knowledge that life may end soon, and yet that it is great to be alive today. Mother wanted to develop my nonexistent artistic abilities, and I went to a professional ballet school run by Mme. Preobrazenskaya for a while. It was clear that I was inept as a dancer. Then came the period when I had a piano teacher who thought I might have talent but who said, bless her, that she was unwilling to continue to teach me unless I practiced four hours a day. Finally, there was a painter friend of one of my cousins who was commissioned to take me to museums and make me art-literate. None of this really took, although I still go to museums—particularly ethnographic museums—in the cities I visit. I had decided to become a pilot and an aeronautical engineer, and I wanted to go to the École Centrale (the MIT equivalent in France) when I finished the *bachot*. This was a way of declaring that I loved Misha, and I reveled in talking to him about math and about scientific questions, which I formulated from reading and from my visits to the Palais de la Découverte. He showed approval when I solved a problem, but he always insisted, usually in vain, that I show him several logical ways of solving the same problem. I was given books and erector sets as presents. I built scale models of airplanes. I never had a doll, though I suppose I could have bought one with my allowance.

There were other tutors: Miss Smith taught me how to speak English by reading Shakespeare and Shaw outloud, yet I never even managed to pronounce correctly the "th" in her name. And there was a succession of German refugee women who were hired by Mother to stay with me in the evenings, and to take me on school vacations to the mountains, particularly to Chamonix and to Argentière in the Alps. The women tended to be overqualified and as bored with me as I was with them. The forced marches over rocks and dales saturated forever my appetite for healthy exercise, as did winter trips to Switzerland where I was supposed to learn how to skate and ski. (I never learned to do either.) At other times, when Misha was not involved in the intense but seasonal work at the Sucrerie, we drove through different parts of France, with me as navigator and Mother a complaining passenger. We stayed in rural inns and in modest hotels; but we ate at excellent restaurants recommended by a book published by the Club des sans-Club, which carried idiosyncratic comments similar to those of Gault-Millau. Except for the winter trips to Switzerland, which I took with Sara, we did not leave France. We had become Polish citizens (since my father was born in Warsaw), and while we were legal residents of France, there was a great deal of unpleasant red tape involved in obtaining visas. The Polish government was extremely anti-Semitic, as indeed were many French, and most functionaries did not facilitate the passage of Jews.

It was in Moissy-Cramayel, in August 1939, that we heard on the radio about the Nazi-Soviet pact and of the subsequent invasion of Poland. On the shortwave set we heard Chopin being played on the Warsaw radio until the music stopped, and we heard the rantings of Hitler and the "Sieg Heils" of hundreds of thousands of Germans. And World War II started, as we knew it would; and then time appeared to stop. In the west, there were no military activities. It was *la drôle de guerre* (the phony war). Men were called into the army and practiced their maneuvers, but it was not clear for what. Yvette had gotten married in 1937 to a French engineer of Polish extraction. He was a lieutenant in the French Air Force and stationed in Dijon. She was able to follow him there. We lived in Moissy to be near Misha and visited Isaac in his warrenlike apartment on Avenue Marceau to hear the latest political news.

By early 1940, we were again living primarily in our apartment in Paris. The mood in France was unnerving. There were extremely strong class divisions. Much of the upper-class "nobility," which was very Catholic, was also fascist. But they were not Nazis: to be followers of a former house painter would have been a tad declassé. Still, they were anti-Semitic to the core, reenforced in their beliefs by the attitude of the Catholic church. Many Frenchmen had died in World War I, only a generation before, but since so many of the officer class had fallen on the *champs d'honneur* (the fields of honor, and of glory), the upper class felt that their nationalistic views were vindicated forever. They were vehemently anti-British, due, I suppose, to pre-twentieth-century history and a subliminal awareness of the decreased importance of France in this Anglo-Saxon century. And they were scared by the inroads of social democracy. In 1936 a prime minister, Léon Blum, had been elected, and a great wave of strikes hit France. I remember that rumors circulated everywhere that workers would take over the factories and massacre their managers and owners, and their families, on July 14. In retrospect this was idiotic, but I did spend a sleepless night in Moissy. But on July 14 I was relieved when the head of the union at Lieusaint, also the local head of the Parti Communiste, came to our door with a French flag and a large bouquet of flowers. Misha had long before instituted the forty-hour week and paid vacations, which were two of the most important gains of the 1936 strikes. Thus, in 1939, the workers were deeply opposed to the war, not only for the usual reasons that make all sane people oppose war, but because a majority of them were either members of the Parti Communiste or in sympathy with it. They did not understand that the Nazi-Soviet pact was a cynical decision by two dictators to carve out some territory for themselves while getting ready to obliterate each other. That Stalin was incompetent at anything but the slaughter of his own people was not understood in 1939. It was not until 1941, when Russia was invaded by Germany, that the Communist groups in the Resistance became really active, although some of the intellectuals in the PC were marginally involved earlier. And in the thirties, France had an additional and devastating problem: its press and its politicians, with rare exceptions, were corrupt, and known to be so. In addition, a rampant bureaucracy greatly impeded daily life. When I visited the Soviet Union later in life and coped with the bureaucracy there, I had a sense of déjà-vu, though the food in France was better.

In 1940 Misha was fifty years old. As the director of Lieusaint, and as a general troubleshooter, he owned one-tenth of the family business which Isaac had started in France. In addition to the sugar refinery, this included alcohol distilleries in Normandy and Brittany, at Les Andelys, St. Lo, Redon, and at other places. Isaac had a daughter and four sons (one of whom was confined to an institution). He was the patriarch of the family and held frequent gatherings for all of us. The ones on the High Holidays were command performances and included us, all my aunts, my other uncle, and a variety of visiting Zionists. I was always the youngest person there, and I was extremely resentful about being forced to attend. I loved Isaac and Sara, but I did not care for Isaac's wife (whom he had married in an arranged marriage), and I didn't much like my cousins either. I suppose that Isaac's wife was a good hostess and mother, but I thought she was a very stupid woman: for example, she welcomed Isaac's concubine into her home. In retrospect, perhaps she did this out of love for Isaac and an understanding of his needs. When Isaac died many years later and his children wanted to discontinue the concubine's monthly allowance, she insisted that it continue. In any case, the rituals of the High Holidays bored and irritated me. I don't think that Isaac was religious, but he was profoundly a Jew and felt that it was necessary to affirm our background. My parents were agnostic although they too were culturally Jews.

As for me, I have been an atheist since the age of ten. Since 1936 it has been impossible for me to believe that a God exists who cares about humans. If he doesn't care about us, why do we need to acknowledge him? Moral principles, dependent on the common-sense golden rules of social interactions, but independent of religion, are far more meaningful to me. I am very conscious of being a Jew and, of course, I respect people who do have religious beliefs. I am wary of organized religious groups, and I still have an antipathy toward the Catholic church. I have not forgiven them for the anti-Semitism they condoned.

Misha's relatives were not a part of my childhood. One of his brothers had died early in Poland, my aunt Luba lived in Warsaw, and two other brothers and a sister had moved to Palestine. His sister Vera Doljanski was married to a physician who was killed in an Arab ambush of a Red Cross truck. She had worked with him in cancer research, and my most vivid memory of her is of the time she came to visit us, bringing cages of white mice or rats that she had obtained in France and was taking to Palestine. Mother and Vera cordially loathed each other. It was difficult to know exactly why, if indeed they had any reasons. Both were strong-willed women who took a proprietary interest in Misha.

Luba was killed in the Holocaust. Even though she hadn't visited us in France, she somehow had established a bank account in England with a deposit of 1000 British pounds and had willed it all to me. I received the money after the war. And a great-aunt who, I was told, had lived like a miser in Eastern Europe, had bought several gold bar brooches, with small, old-fashioned diamond settings. She sent them to several female relatives, with the instruction to hold the brooch for an emergency and then to hock it. I still have my brooch.

On May 1, 1940, I woke up with severe abdominal cramps. Our surgeon, Henri Mondor, diagnosed appendicitis and operated immediately. On May 10, the Germans invaded Holland and Belgium and, shortly thereafter, France. Our family physician told us that friends in the Deuxième Bureau (one of the French intelligence services) had told him that the French and British armies would not stop the attacks, and he advised us to leave. Mother, Yvette, and I went to stay in Brittany, in La Baule, which we had often visited on vacations and because it was close to one of the distilleries, in Redon. La Baule has one of the best beaches in France. It was then a quiet and pleasant resort, and our rented apartment was in the one tall building in town, on the Avenue Pavie, about a third of a mile from the beach.

The news became dreadful. British troops were caught in the debacle and a quarter of a million men were evacuated from Dunkerque. Late in May, to our immense joy and relief, Misha joined us at La Baule. He arrived in his car with some additional suitcases, some money (in retrospect I think it must have been the equivalent of $25,000), and, most important, a set of documents and ink stamps labeled "Ministère de la Guerre." Lieusaint, as was the case with most factories, had been requisitioned by the War Ministry, and Misha had continued to run it for the war effort.

The Germans were advancing, and we wanted to be as far away from them as possible. We still hoped that they would somehow be stopped, and so it appeared to make sense to go south, though the immediate problem was that gasoline was impossible to obtain. In case it became necessary, Misha showed us how to commit suicide. Mother's pills would take too long, so he gave each of us a dagger and told us how to cut our wrists. He was very clear about the advantage of dying by our own hands, with dignity, rather than be killed by the Germans. Shortly thereafter came the first bombing attack by German planes. We took refuge, with other people, in the underground garage of our apartment house, though it was not a particularly safe place. We heard the bombs falling through the night, and later realized that it was the port of St. Nazaire, ten miles away, which was being targeted. In the early hours of the dawn, I could no longer stay cooped up in the garage, and I went outside and walked for a while. I met two British soldiers, and we talked. I spoke some English, because of Miss Smith, and I explained our problem to them: we needed gasoline. Did they know of any place where we could get some? And those two wonderful people said, sure, they were about to embark for England, and they could transfer gas from their army vehicles. They soon filled our tank, and they gave us additional gas cans which we strapped to the outside of the car, along with my bicycle. We offered them money and they refused it. They went to St. Nazaire. If they did embark, they were probably killed, because the Germans blasted the British ships. Though we never even knew their last names or addresses, these men saved our lives.

We packed up and started south along the coastal road, as did many others. The roads were clogged, people were in a state of panic, and once we were strafed by German fighter bombers. Finally one evening, we found ourselves at the end of the coastal road, in Royan. We had to decide what to do next. We could either go southeast toward Bordeaux, on the same bank of the River Gironde as is Royan, or try to cross the estuary of the Gironde by car ferry, heading due south. We tried to sleep in the car, and once again I was restless and I wandered around the many other cars parked along the beach road. And once again I had an enormous stroke of luck: I overheard some Polish and British officers talking about the alternatives. They had heard that the French police were detaining all non-French nationals in Bordeaux, at road blocks. We decided to cross the estuary and did so, without incident, after a night of bombardment by German planes. And once again we were lucky. We later found out that the Germans had mined the estuary during that overnight attack, but our ferry somehow slipped through. Finally we stopped in a small town in Gascony, and tried to figure out what to do next. We could not go much farther south: soon we would come to the frontier with Spain which, while not technically in the war, was totally on the side of the Nazis and, in any case, we did not have visas to enter Spain.

We were all extremely depressed. Mother worried about her relatives and Yvette about her husband, who was somewhere with the French Air Force, if indeed it still existed. As for me, I just wanted to survive. On June 22, 1940, Pétain announced the terms of an armistice he had concluded with the Germans. They would occupy part of France, including the coastal zone in which we were located, and part of France would remain under the nominal rule of the Pétain government in Vichy. He also announced that one of the details of the agreement was that no one in the zone to be occupied could cross over to the Vichy zone. Something had to be done, and Misha did it. He decided that we should go to Toulouse, the nearest large town in the Vichy zone, and he forged appropriate-looking papers for us, based on the War Ministry documents and the ink stamps he had taken with him from Lieusaint. The forged papers were good enough to get us through the road blocks that had immediately been installed by the French. We made it to Toulouse and luckily got a hotel room, in a town swamped with refugees. In the evening there was a mob scene with drunkenness and hooliganism. The French were intensely joyous that the war had ended, for them. It was unreal. The French had forgotten their ethos and their history, and most of all their *pudeur*, a sense of how to respond to disaster in an appropriate fashion.

The question we faced was what to do next. Food was scarce and the sanitary conditions were appalling. Mother was quite ill, so I would stand in food lines to get some rice. My hair was full of lice. We also wondered where Yvette's husband, Ziutek, was. One evening, as we were walking along the streets of Toulouse, I heard a sudden scream of joy: he was here! Ziutek had pulled back with his unit, which was stationed temporarily near Toulouse. Months earlier, he and Yvette had decided that if they were ever separated, they would try to make it to Toulouse, and he was indeed here looking for her. And then we were lucky again. A Dutch consular official decided to save a few Jewish lives. He gave us visas to go to Curaçao, which belonged to Holland. He explained that the visas were meaningless, that we could not really go there, but that, perhaps on the basis of these visas, we could obtain transit visas from Spain and Portugal, to try to get as far away from the Germans as possible.

There were two problems: the first was that the nearest consulates were in Marseille; and the second was that Ziutek could not leave France because he had not been demobilized, and Yvette would not leave him. It was a very difficult decision. Misha, Mother, and I decided to try to leave France. If we succeeded, we would be in a position to help Yvette and her husband. So we traveled on to Marseille and stayed in a hotel near the railroad station. Every day Misha went to the various consulates and waited. Food, other than fish, was scarce. We had fish soup three times a day, which was not a hardship in Marseille. The city was alive with rumors, very few of them optimistic. Nothing seemed to be happening. We were helpless. Then, finally, Spain and Portugal gave us transit visas. The next step was to obtain an exit visa from France. Eventually, to our joy, that too was granted. We prepared to leave, selling the car, my bicycle and anything else we could dispose of for cash. We were among the very few people who had been lucky enough to obtain legal exit permits, and Misha was contacted by some kind of American and British intelligence agents who asked him to take some papers and some information to Madrid and Lisbon. Mother and I thought that they were probably provocateurs, but my father insisted that we must try to help. French regulations of the time precluded our taking valuables with us, particularly currency, yet we needed some to survive in Portugal. Since I had a photographic memory (which grew dimmer with the years), I memorized some of the information to be delivered in Portugal (they were lists of locations and times of illegal crossings over the Pyrenées) and Misha secreted the written material and currency in our suitcases.

In October 1940, the Pétain Vichy government interned all foreign Jews in "unoccupied" France in camps; later they were deported to Germany and most were killed. We had left France at the end of August 1940, taking a train along the Mediterranean coast through Perpignan to the Spanish frontier. On the French side there was no trouble. On the Spanish side, however, it appeared that the guards suspected we were carrying something "illegal": we were hauled off the train, our luggage was thoroughly checked, and we were thrown into jail—after my mother and I first had to endure a disgusting check of our orifices by a dirty female guard. I still remember how scared I was. I promised myself that I would never, ever again try to sneak anything whatsoever through customs, assuming that we would eventually make it through, and I never have. But we were lucky. The guards did not find the written material and the currency in the luggage. Misha had been creative once again. (I will not explain how he did it, on the off chance that it might be useful in the future.) We resumed our trip. I found the Pyrenées and the Spanish landscape beautiful, as was Barcelona, where we spent one night. Then we took the train to Madrid, where Misha delivered the first part of the secret papers. While he was away from our hotel, Mother and I were shriveled up with fear. But all went well, and the next day we arrived in Lisbon, a beautiful city by the sea. We stayed in a small hotel on the principal square, the Rocio, and wandered around the city overwhelmed by the quantities of food displayed everywhere.

We stayed in Lisbon for about three months. Lisbon was the edge of the world. It was filled with refugees from Nazi-occupied Europe. From Lisbon one could escape by ship to safety. Some of the refugees were Britons and Americans: they could return home, but space on ships was extremely difficult to obtain. For European Jews the situation was worse: they also needed visas to settle somewhere. Misha tried to get permission for us to go to England. The Battle of Britain was going on, and England would not take on additional refugees. Over the weeks, Misha explored all possibilities. The U.S. consulate was reputed to be filled with anti-Semites. In any case, the number of applicants was huge and the requirements were daunting. The residues of the Great Depression and the isolationist movement in the United States acted against the admission of refugees, though Eleanor Roosevelt used her influence to allow prominent intellectual leaders to enter. The American to whom Misha delivered the last of the Marseille papers was apologetic but candid. It would be impossible to emigrate to the United States.

Gian Carlo Menotti wrote an Opera called *The Consul* about those times. When I saw it in New York in 1952, I wept. It was a fair picture of Lisbon at the time, with consular offices filled with despairing refugees. Not only did they need to find a place in which to settle, but also some means by which to survive in Lisbon. The long-range problem for many refugees, and certainly for us, was how to survive on the funds we had. I doubt that by then we had more than $10,000 (in current dollars) for the three of us, and there was simply no way of getting a job in Portugal. The short-range problem was that we all believed that Portugal would be invaded by Germany and that we would die. The dictator of Portugal was a man named Antonio Salazar. He, and the Portuguese, had traditionally maintained close ties with Britain, and they did not like Spain and the Nazis, though they realistically accepted the presence of known Nazi agents. We imagined that Hitler did not need to accept a pro-British pimple on "his" continent and that he would occupy it, being granted safe transport by Franco through Spain. (It is true that Switzerland was also neutral, but, to put it as charitably as possible, it was compliant to Germany.) The scenario was realistic, and Britain made arrangements for ships to remove Allied nationals from Portugal if an invasion did take place. We had to get on an evacuation list, and since we were Polish citizens, it had to be that for Polish nationals. It was then a little over a year since Poland had been conquered, but there was still a Polish consulate staffed by the Polish government-in-exile. The Poles refused to put on the evacuation lists for the British ships the names of any Polish Jews. This was extremely depressing and, once again, we went over our plans for suicide if the Germans were to march in.

And then, my roaming around and my speaking English helped us again. I met a British couple, waiting to return to Britain. The gentleman had been a minor intelligence officer in Italy. They were nice people. I told them about the Polish evacuation list, and somehow they managed to place our names on the evacuation list for British citizens. This pulled us out of our depression, and Misha once again made the round of the consulates. We never saw these two people again, but we corresponded with them for a number of years. And, finally, there was a breakthrough for us and for a number of other people. The dictator of the Dominican Republic, Rafael Trujillo, decided that skilled technicians and professionals would be useful to his country. He had his consulate issue immigration visas for such people, and we received permission to go to Santo Domingo. Misha would be able to work there.

The next problem was how to get to Santo Domingo. There was only one way to go, via New York. Going via New York required a transit visa and space on a ship. In the halls of the American consulate were crowds of pathetic and patient refugees. But by now Misha had lost his patience. He entered the consulate with the Dominican immigration visa and demanded a transit visa. He managed to convince the consular officials that yes, dammit, he was going to go to Santo Domingo, and he just needed to go through New York. He meant it, and after all that he had been through, from Russia on out, he was not going to be stopped by some third-level clerk. I suppose that because he was demanding and tough, rather than cowed, he became acceptable to the official.

The last hurdle was to get space on a ship. Few ships were departing Portugal. The shipping line of choice was the American Export Line, whose ships, from still neutral America, were considered safe. Space on them was usually obtainable only by paying huge bribes, but we had very little money left by November 1940, and Misha was opposed to paying bribes. But we did get tickets: Misha walked into the company offices, loudly demanding tickets at the official price, and expressing forcefully his disgust at the bribe taking. It must have been a wonderful scene, and it made the rounds of the refugee circles. In fact, Mother and I even got a proper cabin though Misha traveled steerage. Many years later I found out that the man who was to become my mentor and best friend, Tom Lauritsen of Caltech, who had been in Denmark spending a year at the Bohr Institute, also traveled on that ship in steerage with his Danish wife.

Throughout our three-month stay in Portugal, we had grieved for our relatives still in Nazi Europe, about whose fate we knew nothing. Yvette and her husband were constantly on our minds. But I remember also long walks with Misha through the beautiful city of Lisbon, to the hill above it, to the harbor and quays, through the narrow streets around the main square and to the lush suburbs. We left Europe around December 1, 1940, and we were enormously happy to go. I remember little of our ocean trip: I was seasick from the time we left until we approached New York.[2]

2

America!

1940–1952 · · · · · · · · · · · · · ·

I HAVE NOW LIVED IN THE UNITED STATES for fifty-two years, and I have been an American citizen for forty-six of them. I consider what defines me: what things are the most important to me? And I try, probably foolishly, to rank-order them.

I am an American

I am Wally's wife

I am a teacher

I am a nuclear physicist

My friends are Denys, Ed, Frank, Gloria,

Heinz, Margie, and Yvette

I am a woman

I have cancer

I am a Jew

When we saw New York for the first time in December 1940, my parents were fifty and I was fifteen. The skyline of the city, seen from the ship, was magically beautiful, but we were not allowed to land. The immigration officer did not believe that we would just transit through New York and wanted someone to guarantee that we would leave. We were put on a tender and taken to Ellis Island, where Mother and I were separated from Misha and placed in a cell with a dozen other women

for the night. The windows were barred, and the door
was electrically controlled. During the daylight hours
we were allowed to move about a warehouse-like room.
The place smelled of disinfectant and depression. Many
of the people there had been detained for months. As
newcomers, we were informed of every possible unpleasant contingency by our cellmates. The guards were not
unpleasant but for them it was just a job, and their English sounded harsh and was almost incomprehensible
to me. Had we known that our stay on Ellis Island would
be for just one day, it would have been a trivial, possibly
interesting, experience. But we did not know what would
happen, and that made it a very long day, and it jolted
us badly. Ellis Island may be a romantic place to visit for
the children of immigrants, but I have never had the
desire to see it again, nor do I have particularly tender
feelings toward the Statue of Liberty, which I could see
through the windows of our cell.

The person who got us out was Isaac. He had been
able to escape before us, though my aunts and my other
uncle had remained in France. He was no longer wealthy,
but he had known a number of American Jews through
his Zionist activities and they had helped him. He guaranteed that we would leave the United States and picked
us up at the dock. We landed in New York with $100
(probably the equivalent of $1,000 now). He moved us
into a tiny apartment at the Hotel Hargrave, at 72nd
Street and Columbus Avenue. New York was like a
dream city. It was complicated but perfect. My memory
recalls it as clean and exciting. There were no problems
with muggers or with drugs. Later, when I was sixteen
years old and an air-raid warden (I was so innocent, and
so proud of being a warden), I confidently patrolled 73rd
Street between Columbus Avenue and Broadway at night.

Now I would be reluctant to walk it even in the daytime. My euphoria, despite our anguish about our family in Europe, was partly the result of Mother's attitude. She had become constructive and reasonable since the time we left La Baule. She was no longer bored. She no longer screamed demands. She coped well with the day-to-day difficulties. My parents' marriage strengthened, and I was happy.

We had planned to go on to Santo Domingo but we instantly loved the United States and wished to remain. I wanted to become an engineer, and this would surely be easier here. Mother wanted to remain in an urban area and be near Isaac. I don't know what Misha thought, but I suppose that above all he wished to ensure my future, knowing he himself could have a satisfying life anywhere. He began to inquire how we could change our status from visitors to immigrants. It turned out that one could not do this while residing in the United States. We had to leave the country and apply to a U.S. consulate. An American consul had a great deal of leeway in deciding whether or not to award immigration visas. Officials in Canada were reputed to be sticky. The consulate in Cuba was thought to be more reasonable. After two months in New York, we sailed to Cuba.

Our arrival was rather unpleasant. Though we were fastidious Europeans, we were assumed by the immigration authorities to have exotic diseases (I remember that Mother and I were tested for syphilis). We were allowed into Havana and stayed in a small hotel in the center of town. We immediately applied for immigration visas to the United States, and after a few weeks they were granted to us. We benefited from a very helpful loophole: there were quotas for immigrants from different countries. The Polish quota was full but the German quota was open, because few Germans were able to leave. By German law, when I was born in Berlin I was assigned the nationality of my parents and I became a Pole. For the purposes of the American quota regulations, however, I belonged to the German quota. I could get in, and my parents could then get in because they were my family.

In the meantime, Misha and I wandered around Havana and the Veradero Beach area, where some very distant relatives lived. It was an interesting place, but I yearned to be back in New York, where I had already started the tenth grade. I still had a lot of difficulty speaking and even reading English, and in Havana I tried to pick up some Spanish as well, because I was curious about life in Cuba. Since we had virtually no money I could no longer buy French books, but there was an excellent American library and I began to go there daily during our stay. It was a marvelous place to begin to understand the United States: I read magazines and books voraciously. And finally in April 1941 we sailed back to New York.

When we had decided to try to stay in the United States, I had called the Board of Education and found out that the high school nearest to where we were staying was Julia Richman High School. I enrolled there. The cultural shock was enormous. Its atmosphere was certainly not as permissive as that of present-day schools, but compared to a French lycée it was extremely relaxed. There were electives and student organizations. On the other hand, the students appeared to be totally unaware of what was happening in the outside world. In a European history class the teacher asked for the name of the prime minister of Britain. A couple of students guessed Chamberlain; no one knew that it was Churchill! The implication was that the war and the threat to the United States was not discussed in the children's homes, which was hard for me to understand. Julia Richman was an all-girls school. The school had within it an enclave, called the Country School, whose students were planning to go on to college, a step considered appropriate before marriage and family. Many of them looked down on the students who took commercial courses. I was not used to that: all students at a lycée were considered to be academically oriented, and there were no second-class citizens. I should add that before World War II, few working-class students were able to attend lycées, but the discrimination was not so obvious on a daily basis as it was at Julia Richman.

I needed the better classes and the academic subjects taught within the Country School, but Misha insisted that I also take typing so that I could earn a living if college did not work out. In France he had insisted as well that my sister Yvette learn to be a secretary. And she indeed earned her living as a secretary in Palestine, in France, and in the United States until she received a Ph.D. and became a professor of Romance languages at Queens College. Yvette and her husband, who were French citizens, were not interned in 1940. (Internment and deportation of French Jews began in 1942.) In 1941 they were able to leave Toulouse and sailed for Cuba from Vigo, a Spanish port. By then we were settled in New York and we wanted desperately to have them join us. It took a year or so to get the State Department to agree. Besides writing petitions, I went to Washington two or three times to plead for their acceptance. It probably helped that I was a very determined sixteen-year-old girl.

I failed English during my first semester at Julia

Richman High School, which had been interrupted by our stay in Cuba, but I did relatively well in my other courses. I started taking the State Regents examinations and did receive high scores on those. I had been advised that they were extremely important for admission to a university. Two of my teachers were incredibly nice to me: Eleanor Wheeler, my English teacher in my junior year, encouraged me to write; Sara Solon, my typing teacher, was an old-fashioned Jewish mother who invited me to stay with her family in their cottage outside New York on school holidays. Both helped to keep me out of trouble. During Brotherhood Week, we were required to write an essay on the glories of brotherhood. I wrote a blistering paper on the hypocrisy of having a Brotherhood Week, regurgitating noble short-lived sentiments and behaving abominably in daily life. I had begun to realize the extent to which racial bigotry dominated the United States at the time. I was told by another teacher that there had been a move to fail me for that paper, but that Eleanor Wheeler insisted that the paper was well written and that this should be the determining factor.

With the exception of the racial problem, which has always bothered me, I loved the United States almost instantly and felt that it would be my country. On the other hand, I was bothered by the daily pledge to the flag. I felt that it was not appropriate for me to pledge when I was not even a citizen, so I chose to stay silent. It was not easy for me, and I suspect that my teachers were also uncomfortable. As I write these words it is July 4, and *my* flag is flying outside my house. I got up at 6 A.M. to put it up. I am proud—and so thankful—to be an American.

My father was having quite a difficult time getting

jobs. His English was not very fluent, it was heavily accented, and he was in his fifties. It was hard for a proud and independent person once again to start over in a new country. He took engineering consulting jobs, one of which was with Armand Hammer, who later became the CEO of Occidental Petroleum. Misha considered him a very unpleasant person and stopped working with him. Paying the rent and making it through the end of every month were very difficult problems for the first few years, but, of course, our main concern was our emotional involvement with our relatives who had stayed in Europe. We had no news until the war ended, but we knew it would be bad. All but two of our relatives were killed. My aunts, Sara and Fania, hidden by Christian friends in Paris for four years, had survived.

At the same time, I was having a glorious time discovering New York City. I traveled all over it, went to all the museums and to the beaches on Long Island, and, when the heat of summer came, my parents and I went to air-conditioned movies, if we had the money. Besides being an air-raid warden, I joined the American Women's Voluntary Services. I proudly wore a uniform as a member of the AWVS, and *Life* and *The Daily Worker* wrote stories about the young girls in this organization, with pictures. In one of the pictures I am shown warily eating a hamburger, my first. After reading countless brochures, I had decided that I wanted to study engineering at MIT, the University of Michigan, or Purdue. All required a subject called "Solid Geometry" which was not given at Julia Richman High School. Two or three of us formed a study group and took the Regents exam in that unlikely subject, and we passed it. So I applied to the three schools: I was accepted by Michigan and Purdue, although without scholarship assistance. As for MIT, I was interviewed by an alumnus in the Wall Street area, who was candid enough to tell me that there was a *numerus clausus* for women and one for Jews, and that my grades were not good enough to overcome both. I went to Michigan.

I graduated from Julia Richman High School in June 1943, but I did not go to the graduation ceremony. Except for Sara Solon and Eleanor Wheeler, I did not have any particular fondness for the school. I wanted to start at Michigan as soon as possible, because the engineering program was five years long. During the war, Michigan had three terms a year, without lengthy vacations. I started at Ann Arbor in June. Uncle Isaac gave me $2,000 toward my college expenses, my parents added as much money as they could, and I still had my great-aunt's gold brooch to hock if necessary. Misha also gave me a foot-long slide rule which I wore dangling from a belt, and I was off. Of course I immediately became extremely lonely for my family.

I spent the summer in a large and impersonal dormitory. And I was the only woman in my engineering class of about one hundred men. The young men were extremely nice to me, and throughout my undergraduate days they treated me as a younger, very naive sister. In fact, the only date I had ever had before Michigan was with the son of a friend of Misha's who had been cajoled to take me out a couple of months before my departure for college. Life was suddenly very different. I did not date that first semester at Michigan, but I did go out drinking with the guys. At least I had been prepared for that by Misha. He drank very little, but he made his own vodka, using very pure alcohol to which he would add a few drops of glycerin and some herbs. Before I left, he took out a bottle of this brew and showed me how to get it down, after telling me always to begin by eating a lot of greasy food. It worked for me at Michigan, where I drank a great deal since it was forbidden, and in meetings with Russian colleagues thereafter. On the whole, though, I much prefer Coca-Cola. My buddies also taught me a lot of vulgar slang, which I was happy to learn and which I continue to use freely.

Michigan has an extraordinary women's dormitory called Martha Cook that was given to the university by a lawyer, an alumnus. It was named after his mother, and he had specified that it should foster a gracious style of living, including good food. The rooms were large and very well furnished. At that time admission to this civilized dormitory was by decision of the director of Martha Cook, the widow of an American ambassador. It was said to depend in part on the applicant's academic record, but she took me in even though my grades had plummeted as soon as I started taking courses. It was some time later that I understood that there was at least an informal quota of Jews: there were never more than two or three Jews at Martha Cook during my three years there, out of some 120 girls. But I felt absolutely no anti-Semitism. If you were admitted, you were in. The atmosphere was extremely friendly, and we were supposed to memorize the names of all the girls at Martha Cook. We had yearly exams on this, and I usually flunked. The young women were bright and very pleasant, and most were preparing to be excellent wives. One of the most intelligent married a man who became a senator, another is a judge, several have become writers, and three of us are college professors. My most notable contribution to Martha Cook was as chairman of the Date Bureau. The only suitable young males in town were in a navy program, the V-12 program in the engineering school, and I knew them all.

The College of Engineering had a number of required courses. These included a course in forging and welding, one in mechanical drawing, a single course in engineering English, and four years of chemistry. In addition, there were perfectly sensible courses in mathematics, physics, and in the engineering subfield in which one planned to work. I loathed the first set of required courses. The welding was not so bad, but heating a piece of metal, lifting a heavy hammer, and then swinging it to hit the metal before it cooled was beyond my abilities. A lot of nice people helped me to barely complete the requirements. The mechanical drawing course was hardly better: I had a terrible time visualizing things three-dimensionally. I can still make quite excellent two-dimensional ink drawings, using my leftover Leroy lettering set, but now, of course, all these neat things are done by computer.

The engineering English course was something of a joke. Were we considered too dumb to take a regular English course? We were taught by a sad little man hired by the engineering school who decided that we should read one book during the semester. He chose *Tono Bungay* by H. G. Wells to represent the best of world literature. Consider yourself lucky if you have never read this book. We were supposed to write an essay about it, and I did—stating exactly what I thought of it and of the course. At the time I was still naive enough to believe that one should state one's views candidly and strongly. Quite reasonably, this did not endear me to my instructor. There was talk of my failing the course but I was rescued by some essays I had written at Julia Richman High School under Miss Wheeler. I entered them in a writing contest for undergraduate and graduate students at Michigan and won an Avery Hopwood Prize. The $75 I received was very nice, and the prize brought much pleasure to me, to my parents, and to Miss Wheeler. And it quite directly contributed to my passing engineering English.

As for chemistry, it was a disaster. For some reason I disliked the subject immensely, and I was particularly inept at laboratory manipulations. While I had a very good, almost photographic memory, I hated to memorize chemical formulas. This was probably a leftover rebellion from the French educational system in which we had to memorize everything from English words, to poetry, to the names of French kings and dates important in French history. In any case, my ability in chemistry was close to nil. I remember two incidents: in the qualitative analysis course, each of us was given a different chunk of material, and to pass the course we were required to determine its composition. The first step was to dissolve it in an acid. The kind of acid which accomplishes this is the first link in determining the answer. But I couldn't dissolve that clump. I tried every acid available, and every combination of acids. It still would not dissolve. The end of the semester was approaching quickly, so I finally decided to try to solve the problem creatively. I had taken a course on engineering materials and compared the lump's appearance with all the listings in the engineering book. I took a chance on it being "Wood's alloy," and wrote down its elemental composition on my answer sheet. I had guessed correctly!

The second incident occurred during the spring semester of my senior year. I had the good fortune to get a really bad case of measles followed by a very severe allergic reaction to the antibiotic I had been given. This prevented my failing physical chemistry that semester, at the expense of having to stay an extra summer. Ironically, it was my minuscule knowledge of chemistry which led me to solve a physics research problem and receive a Ph.D.

I had entered Michigan intending to become an aeronautical engineer and a pilot. It was quickly and candidly brought to my attention that as a woman, the best job I could hope for was as a glorified draftsman, and that I was a lousy draftsman. In the meantime I took flying lessons. My instructor, at a small nearby field, was puzzled by the fact that I wanted to fly but did not even know how to drive a car. He was forced to modify his teaching to account for the fact that I had no knowledge pertinent to flying, and I had a great time for the ten hours my log book shows that I took lessons. I suspect that he was dismayed that the only way I could cope with the stall practices was to close my eyes and count to ten before gently raising the nose of the plane. To get even with me, he would direct me to fly around, and then ask, "Where is Willow Run Airport?" and I would try to locate it, and tilt the wings this way and that, and only after a while did I realize that he asked the question only when the airport was directly under us. I was ready to solo, but I was under age and had to get my parents' permission, which they refused. I think that I was not overly unhappy about this, since it was clear that I had no talent whatsoever for flying. But its glamor was still there. I joined the Civil Air Patrol Cadets, which gave me a uniform, a chance to take ROTC courses and, for some reason, to learn navigation. The ROTC course I liked best was shooting. I learned to shoot Garlands and .22s. The Garlands had long stocks that dislocated my shoulder when recoiling, but I liked shooting and became quite good at it. I still like guns. I tried to join the Women's Army Corps, but the WACs had a twenty-one-year age requirement. One also had to be a

U.S. citizen and pass a quite strict physical exam, and I was encouraged to forget it.

I very soon switched my major to electrical engineering. Misha had started a factory that produced motor-generator sets, and I knew that I could work for him. But the more I thought about it, the less appealing that prospect became. I was neither interested in nor doing well in the EE courses. I proudly joined the American Institute of Electrical Engineers and attended the local meetings, but I really found the courses very dull. And then I switched to engineering physics. It was a physics-major program, but the fact that it was located within the engineering school meant that I wouldn't lose most of the engineering credits I had accumulated and could graduate in 1946. We still had financial problems and I had to finish as soon as possible.

I became interested in physics for two reasons. One of them was a physics professor, Floyd Firestone, who really appeared to enjoy what he was teaching, in sharp contrast to most of my other instructors. I remember him, a roly-poly man sitting on a rotating stool, demonstrating the conservation of angular momentum, and giggling. The other reason was Marie Curie: she was evidence that a woman was not inherently disqualified from studying the physical sciences. I started taking too many physics courses too quickly and did very badly in them. But I was hooked. I worked for three of the physics professors at various times. I took data for an old gentleman, H. M. Randall, for whom the main physics building is named. He was very pleasant but the work was incredibly dull. I did X-ray crystallography with Professor George Lindsay, whose retirement may have been precipitated by my breaking most of his (glass) equipment in the dark. He taught me to use libraries. Since I knew some German, he suggested I translate a chapter of M. Siegbahn's work on X-ray spectroscopy. I began to do this, painfully, until the day when, browsing through the library, I discovered that the book had already been translated into English by Lindsay. In addition, Professor H. R. Crane allowed me to putter around his electronics lab. I had come across the Radio Relay Book which gave circuits of standard receivers and transmitters, and I wanted to learn to build some simple equipment. There was no difficulty in testing whether a receiver worked, but I had not considered carefully how to check my putative transmitter. I was mulling over the problem when two hard-eyed characters walked in and told me that I was disrupting airplane traffic. This could have been a trick played by my chums in the lab, but it succeeded in turning off my interest in any further hands-on experience at that stage.

The work was hard. During one semester I had twenty-seven hours of classes because of labs and the condensed schedule of the war years. And the signals I was getting from my grades clearly warned me that I was not a born scientist. But I had fun, and I was determined to keep trying. Both my parents insisted that I return to New York for one year after graduating from Michigan. Mother wanted me to apply to the Pulitzer School of Journalism at Columbia, on the basis of the Hopwood Prize and an interview I had had with an editor at Random House. But I wanted to go to graduate school in physics at Columbia. I took the Graduate Record Examinations. My GRE scores were extremely high overall, but in physics they were the lowest possible. I think I received a 200, placing me at the very bottom of all the physics majors in the country. Fortunately, I was rejected by the Pulitzer School, but the physics department had no objections to my taking physics courses, though not as a regular graduate student.

Michigan had been both fun and quite lonely. My friends at the Martha Cook dormitory were wonderful young women but they did not share my interests. During a couple of Christmas recesses I stayed at the Michigan League, a women's student activities building with a few guest rooms, because Martha Cook was closed for the holidays. I planned every day very carefully to give myself some special treat: a movie on one day, a good meal on another, a walk to the bookstore the next. Misha sent me small checks for such treats and I bought inexpensive presents for myself, gift-wrapped them, and opened them when I felt really blue. I had also started to go out on dates.

As at most universities, special activities were available for foreign students. I was by then thoroughly assimilated into my American environment, but I liked to speak French (and Russian). In 1944 a group of medical doctors from Haiti had received fellowships from the U.S. government to take public health courses at Michigan. They spoke French and I became friends with them. Among them one person stood out. He was in his thirties, a gentle and very intelligent and cultured man, and we enjoyed talking to each other. His English was very poor and he had difficulties with the courses he was taking. We used to try to find a place that would serve us coffee but were often rebuffed. When I brought him to Martha Cook I was taken to task by the director and by my friends. The man was black, and his name was François Duvalier. I have tried to understand how this kind and idealistic man who, when I knew him, cared so deeply about his people's misery and about the corruption of Haitian society, could have become Papa Doc. I do not know the answer. Perhaps he became mad when he was confronted by power. I do not believe that I was wrong in 1944–45 in thinking of him as a really good person. Other people who knew him then had a similar view of him.[3] François thought he had fallen in love with me. I could not deal with that, nor with any other close relationship at the time, and I did not answer a letter he wrote me from Puerto Rico the following year. Shortly after Duvalier and several other Haitians left, I was asked, on a local radio show, to explain how very useful it was to the United States to invite foreign students and scholars to come and to learn about the American way of life. I suggested that the money would be better spent on public relations propaganda since racism was such that black foreigners would leave with very xenophobic views, as indeed François did.

I was also interested in a puppy-love kind of way in another man. He was one of my classmates, the son of a professor at Michigan. His nickname was Red, the color of his hair. He was tall, handsome, intelligent, and conventional. He became a member of a fraternity which, at the time, discouraged its members from dating Jews. My sexual inhibitions, and his meeting a very nice young woman whom he married, ended our friendship. He and his parents were kind and warm to me, and I remember them with affection. In retrospect, I also appreciate their not bursting out with laughter when Red took me to the Senior Prom. I was a co-chairman of the Prom Committee, which was nice because I could invite Red as my date and his parents as chaperones. Unfortunately I did not have the money to buy a prom dress, and so I sewed up something vaguely Greek and violently green to wear. I knew it was ghastly even at the time, but I had no choice but to wear it.

In the fall of 1946 I started taking physics courses at Columbia University. In a sense it was a return to childhood. I lived with my parents in a very pleasant apartment at a residential hotel, the Park Royal, on 73rd Street, half a block from Central Park West. At the same time I had dumped myself into a pressure cooker. Columbia at the time had one of the two best graduate physics departments in the world. The other was Chicago. They were attended by the most outstanding students, with great ability and strong backgrounds. Many had worked during the war at the Rad Lab at MIT or on the Manhattan Project. I was totally overwhelmed, and despite much help from my classmates, several of whom are my friends still, I failed the first-year physics courses given by Norman Ramsey, Polykarp Kusch, Willis Lamb, all of whom later received Nobel Prizes, and I. I. Rabi, who already had one. However, being surrounded by such bright people who were on the threshold of discovering fantastically exciting physics made me even more anxious to keep going. I got a job teaching labs part-time at Hunter College in the Bronx during the spring. My supervisor was Rosalyn Yalow, who later received a Nobel Prize in medicine and physiology. And during the spring I took a course in nuclear physics at Columbia given by J. R. Dunning. Dunning asked us all to write a term paper. I had become interested in the cosmic radiation and wrote a long and very good paper. His was the only course I did not fail.

Some years later, after I had become a physicist, I sat with Polykarp Kusch on the terrace of the Shoreham Hotel in Washington, D.C., during a meeting of the American Physical Society. He asked me how I could have done so badly in his course when I was clearly able to be a physicist. The answer is probably complicated: I was unprepared, I was intimidated by the other students' knowledge, I was still growing up, I became unglued at exams. I took several guidance tests at Columbia that were supposed to reveal my strengths and weaknesses. Interpretation of such tests was usually done cautiously, but not in my case: I was told that I did not have any ability for the sciences. In my over forty years of teaching since then, I have taken great care in advising a student who has poor grades in physics. I point out my own case, explain that such cases are very rare, but that if physics is what they *need* to do, they might well continue trying—that the odds are against them but that they are not nil. And that incredibly hard work, and luck, can beat a lot of odds. But after all these years, I know of fewer than half a dozen physicists who made it despite miserable academic records.

Why did I want to become a scientist? I try to recapture my thoughts of nearly fifty years ago. The dominant reasons were that I wanted to earn my father's respect, and that I wanted to be judged objectively. Because of Mother, my cousins, and the students and teachers in nonscientific fields whom I had met at Michigan, I had come to feel that the criteria for saying that a book, or a painting, or a piece of music were good were highly subjective. I had no difficulty in knowing what *I* liked and respected in literature and in the arts, but I did not want to be subject to criticisms that could be personal rather than objective. I thought that scientific work would be appraised logically, and unemotionally. I wanted to avoid emotional outbursts which I related to the behavior of artists (and of Mother), and not of scientists. (I have learned better!) I also had more valid reasons for choosing science: I was entranced by the process of scientific discovery, by the questions one could ask (and for which one could hope to obtain answers), and most of all by the minds of the first-rate physicists I was beginning to meet. I was a physics groupie.

But what gave me the nerve to go on in science? My grades at Michigan had barely been at a C-plus level. My Graduate Record Exam score in physics had been dreadful. I had failed four courses in physics at Columbia. Yet I had learned a number of very useful lessons as a child. Among them was that failing at something was better than not trying to do it. I knew that I had been lucky in the past. I had learned that I was strong, and that my stubbornness had borne fruit many times before. I knew that I could handle responsibility. And, most importantly, *I had never been told that being curious*—asking questions, seeking answers—*and that being strong was not appropriate for a woman.*

The expectations of Misha, and the example of my aunt Sara and my friend Lida, both independent women, were so much more important in my life than was the influence of my peers. I was too naive and unaware to be caught in sexual traps. I enjoyed being different, as I inherently was because of my European background. The fact that I was a young woman in a group of men was, in that context, a relatively minor difference. Being one of the boys seemed entirely satisfactory to me at the time. My daydreams of the period were the Walter-Mittyish dreams of a young man: becoming a good and respected physicist. Of course, I did need to give and to receive affection, but my dreams were almost totally centered on my work.

My chums at Columbia University were quite wonderful. They helped me with my questions and they took me to lunches and to parties. One of the funniest expeditions was to a local beer joint where Martin Block (later professor of physics at Northwestern) slipped a box of condoms into my handbag. When I picked it up, I said "What's that?" opened the box, and asked if they were balloons. A roar of laughter ensued. The guys no doubt tried to explain to me what they were for, and I was probably polite, but my mind was still blocked from understanding the form and function of sexual organs. I was then twenty-one years old.

It was 1947. World War II had ended a couple of years earlier but the situation in Europe was still chaotic. We had difficulties finding out what had happened to our relatives. The Red Cross was unwilling to help with information about civilians. Finally, we found out that Sara and Fania had survived in Paris, never stepping outside their hiding place for four years. We sent packages, of course, but most of all we wanted to see them. Civilian ship space became available at last during the summer of 1947, and Mother and I booked passage for Southampton, sharing a cabin on the old *Queen Elizabeth* with two other women. One of them festooned our quarters with her extra-large size girdles, which she washed daily, and the other appeared to be free in delivering sexual favors to the men on board. She was not often in the cabin.

We had to go through England because the French ports did not yet handle large ships. We had two steamer trunks full of food and clothing, and Britain was still rationing all goods, so I had to convince customs in Southampton that we were not black marketeers, that we simply wanted to pass through Britain on the way to France. Finally they allowed us to proceed with the trunks. They were put onto the train and then removed in London, where we rested for a couple of days with a family friend, Colya Seltzovsky. He put us on the train to Dover, and I negotiated again with the baggage handlers, and the customs officers in England and then in France. I had been granted U.S. citizenship in December 1946, and this was my first trip with an American passport. I cannot describe how wonderful it felt to be an American. When one becomes a citizen one has to foreswear allegiance to one's previous country. I doubt that there are many people who could do this as honestly as I could.

Finally we made it to Paris with our trunks, and we saw Sara and Fania. They were frail and sad, but we were happy to be reunited. There was no point in trying to gain access to our prewar lodgings. We knew that the apartment had been looted. We did go to see our physician, who had warned us to leave in 1940, to thank him. In his apartment hung a candelabra which Misha had designed for Mother. We did not look to see if he had stolen any of our other things. He was cordial but did not refer to his theft. We decided not to mention it either. It was a small price to pay, though I wish he had given us the option of making him this gift. It was a shock for Mother, and this man is one of several people whom I hold in total contempt.

I wandered all around Paris trying to reconstruct my childhood. I decided to revisit Lieusaint and Isaac's country house in Les Andelys. Wearing a Michigan T-shirt and a rucksack and carrying a small suitcase, I went on to St. Malo, the Mont St. Michel, and La Baule. I then headed south along the 1940 route and visited one of my Michigan professors in Pau. Professor Pawlowski was already an aged gentleman when I had taken his course in aeronautical engineering. He was married to a French-woman and they went back to France after the war to retire.

After J. R. Dunning's course at Columbia I had become fascinated by the subject of cosmic rays. Cosmic rays are primarily extremely energetic protons that originate in the sun, in our galaxy, and even outside our galaxy. It was then necessary to go to high altitudes to study them. I thought it adventurous and romantic. (Many of the studies are now carried out with space probes.) One of the meccas of these studies was the Pic du Midi Observatory in the Pyrenées. I somehow got there by asking for rides. Later I went on to Marseille and to the Côte d'Azur, where we had spent vacations at Ste. Maxime before the war. The coast was as beautiful as ever, but it had been a U.S. landing site and most of the houses were demolished. Then I went on to Chamonix, where I met one of my Michigan friends and where Sara had come to rest. At the top of the Aiguille du Midi, a 12,600-foot peak which dominates the valley, was another cosmic ray observatory, directed by Louis Leprince-Ringuet of the École Polytechnique. I had met Pierre Auger, a very well known French cosmic ray physicist, in the United States and he gave me an introduction to Leprince-Ringuet. He in turn arranged to allow me to visit the observatory.

There was only one small problem. The cable car, which now connects Chamonix and the Aiguille and which then sweeps over the Vallée Blanche to Courmayeur in Italy, had not yet been completed. At that time, there were only open cement buckets that ran over the cables. Physicists who wanted to go to the observatory could be hauled up in the buckets so long as they were willing to sign a statement that they would not hold the government responsible if there was an accident. I decided to go, even though a couple of people had been killed shortly before, when a strong gust of wind toppled the buckets in which they sat. I must have been mad. Being carted up there was more or less okay, even though a queasy feeling came over me whenever I stared at the very jagged rocks below. The observatory was not very interesting, but the view of the Vallée Blanche, a huge sweep of eternal snows, was unforgettable. However, to get back down, I had to step into the open bucket that was swinging gently to and from the makeshift platform. The thought that trying to climb down from the Aiguille was sure to be even worse than what I was about to do convinced me to proceed in this manner. That, and my sense of pride—an American could not show herself to be a coward in front of a group of Frenchmen. On the way down a strong wind did manage to toss my bucket and me around a bit. I started to count in Russian (a method I use for preventing total hysteria) and held on to the sides with all my strength. I think I had to be pried away from it when we landed, but I know for sure that I spent the next few days in a catatonic state in Chamonix.

I then proceeded to go over the mountain pass to Switzerland, to visit Lida and her family. Lida had been a dominant presence in my childhood. During the war her husband, Jean-Pierre Vernant, had been the head of a resistance group near Toulouse. She had worked with him, had a daughter, and contracted tuberculosis. She was in a sanatorium and was as beautiful and interesting as I remembered her. Unfortunately both of them were at the time deeply involved in the Parti Communiste and were virulently anti-American. They called me "petite Bécassine," after a French cartoon character, a very naive and stupid little maid from Brittany. I kept trying to convince them that their view of the United States was almost totally incorrect. I failed.

Finally, upon my return to Paris, I found out that I had been offered a job at a community college in Chicago. I had realized before I left that I needed to learn introductory physics, and that the best way to do it was to try to teach it. I applied for jobs through the American Institute of Physics. Because of the masses of returning veterans, community colleges were springing up everywhere and they needed faculty. I was offered a one-year position at the Navy Pier branch of the University of Illinois, for $2,800, and I happily accepted it.

In 1947 teaching at Navy Pier was an experience. The long, narrow pier had been converted into classroom space, except for a lounge at one end with a beautiful view of Lake Michigan and the Chicago skyline. The mission of this community college was to teach elementary college courses to returning veterans. Eventually it became the University of Illinois, Chicago Circle, with a campus in the city. I taught eighteen contact hours a week to a total of 150 students. During the first semester this teaching also included elementary mathematics because we found out through a diagnostic exam that 70 percent of the students did not know what 4 minus 4 equaled, and 90 percent did not know what would happen if a number were divided by 0; and they knew no trigonometry. I smile when my colleagues now discuss the parlous state of students' knowledge. What the students had was maturity and the desire to get on with their lives, coupled with the joy of no longer being in a war. They worked extremely hard, and it was a pleasure to teach them, and to learn from them. Most of them were older than I. Several took me out to a baseball game when they found out I had never been to one. I found it totally arcane and completely uninteresting, but there was something called a seventh-inning stretch which I remember favorably. I have not been to a game since.

At the time, housing was scarce in Chicago. I had finally found a single room in a rooming house fifty-four blocks from Navy Pier. The landlady was quite unpleasant and the common bathroom none too clean. It cost nearly a third of my salary. I did find a small rent-controlled apartment, but it was in a building from which Jews were excluded as tenants. I was told that the mayor of Chicago lived there. Every morning I took a very long bus ride to a spot a few hundred yards from the beginning of Navy Pier, then I sprinted half a mile or so to my classroom near the other end. My first class was at 8 A.M. Still, I made friends and Misha came to visit from time to time because he had business in town. Occasionally he would invite his business friends to nightclubs and I would go there as his hostess. I developed a lifelong aversion to smutty nightclub comics.

Misha's business was beginning to go well. He offered to support me in graduate school if I could obtain a Master's degree in one year. If I could not do so, then I should give up or do it entirely on my own. While at the Pier, I had been informally attending a seminar on cosmic rays given by Marcel Schein at the University of Chicago. There was no question that I did not have the background and the ability to go to graduate school there. (My future husband, whom I did not know then, was a graduate student at Chicago. Four of his classmates have since been awarded Nobel Prizes: O. Chamberlain, T. D. Lee, J. Steinberger, and C. N. Yang.) I liked the Midwest, and I decided that the University of Wisconsin would be the right place for me to attend. Unfortunately I could not meet its nominal entrance requirements. Still, I went to visit Madison in the spring of 1948, had an interview with the chairman, Ragnar Rollefson, and that wonderful man admitted me as a graduate student, though without financial aid.

During the summer of 1948, Mae Driscoll, a colleague at Navy Pier who continued there in the fall, came with me to Madison. We decided to share a double room in a dormitory. She was a very nice young woman but I was totally unbearable. I had never lived in the same room with anyone, except with mother, for more than a few days, and it did not work out with Mae. For many years I thought that I would never be able to live with anyone in close quarters, but I have had no such trouble with Wally. As much as anything, it was a matter of restlessness and the need to be alone. Until Wally, whenever my parents wanted to squelch a romantic relationship, they would warmly invite the two of us home, and leave us alone. It could have been called the four-hour test. At the end of four continuous hours, I was bored to tears, and whatever romantic possibilities had existed, vanished. I suspect that they did this very deliberately. Neither wanted me to get married. And when I did get married, neither wanted me to have children. This is hardly surprising. They had begun their life very much in love, but even though they cared for each other, and behaved in quite responsible ways, their characters and interests were so different that it was a burden for them to live together. In the fifties, Misha moved to New Jersey while Mother remained in Manhattan. They visited each other quite often, rather formally, as guests.

I did adequately in my courses that first year at Wisconsin and even passed the Master's exam at the end of the first year. The first time that I met one of my best friends, Heinz Barschall, was shortly following that exam.

He was professor of physics there and he was feared by
almost all the graduate students because he was very
austere looking. He never smiled. He saw me in a corridor and asked, "Are you happy that you passed?" Col-
lapsing with horror and fear, I said, "Yes, my god, what
happened?" Heinz then said, "Some people don't know
how much stupider other people are." I should say that
Heinz doesn't remember that he made that remark. In
fact, Heinz is one of the kindest and warmest men I have
ever known. At the time he was quite shy. That, and his
formidable ability as a physicist, combined to make him
an awesome figure to most of us. I say most of us because one of his graduate students was not awed by
anyone. He was clearly the best in our group. His name
is Robert Adair, and he has since had a distinguished
career as a particle physicist and as a leader of American
physics.

Bob had been a sergeant in Patton's army and had
been badly wounded in Europe. He is a superb physicist
and even at Wisconsin he was a self-confident man. I
admired him tremendously and I wished that I could
acquire his intuitive understanding of physics. I was still
a groupie. I was not yet a physicist. Bob helped me
through my years as a graduate student by allocating to
me seven be-kind-to-Fay days a year. When I was very
depressed, I could request a BKTF day. Bob would then
feed me lunch, listen to me, and explain all the physics
problems that stumped me, and there were a lot of them.
Most importantly, he would refrain from being candid,
and therefore cutting, to me during that day. He told me
that I had to husband them: more than seven BKTF days
would drive him nuts.

There were no cosmic ray physicists at Wisconsin, but
I was still enchanted by the subject. I had read most

everything written about it, and I was moved by a lyrical account of cosmic radiation published in France by Louis Leprince-Ringuet. I translated the book into English and it was eventually printed in the States in 1950. In retrospect, I think I did a disservice to him and to the readers of the book: my English was not good enough to convey its charm. The summer of 1949 I went back to Europe, to see Sara and to visit cosmic ray laboratories, including Cecil Powell's lab at Bristol, the Birkbeck lab in London, and Louis Leprince-Ringuet at the École Polytechnique in Paris. The latter allowed me to participate in the Great Balloon Race of 1949. Cosmic rays come from outer space and there are more primary rays the higher one goes. At the time, one could observe them best by placing special photographic emulsions (plates) in balloons or in mountain observatories such as the one at the Aiguille du Midi.

On July 14, 1949, we met at dawn in a military airfield near Paris, filled a bunch of balloons with helium, and tied them to a package of emulsions and to a small radio transmitter. The scenario was that the emulsions would get exposed to cosmic rays at high altitude for a few hours, that we would follow the balloons, and then retrieve the emulsions when the balloons deflated. We had a War Department truck with a high antenna and a receiver manned by uniformed soldiers, and an open Jeep filled with Leprince-Ringuet and his assistants, with me sitting on a back ledge. We followed the balloons hither and yon. We often lost them, because it was July 14 and every village had huge banners across the streets celebrating Bastille Day. The truck's antenna plucked off the banners, causing the villagers great dismay. We would stop, try to remove the banners from the antenna, and search frantically for the elusive balloons, which meanwhile had continued to move freely in an unknown direction. I think that the whole exercise would have worked out, but at noon everything stopped and we repaired to a two-hour lunch, with wine. I don't know if the emulsions were ever recovered by some puzzled farmer and returned to Leprince-Ringuet (whose address was tagged to the package), but I do remember returning in the jeep to Paris the evening of July 14, tired out, my face and hair black with soot, suffering a major hangover. The remainder of Paris was celebrating. I crawled into a hot tub with an ice bag on my head.

Mother had come over to Europe with me, and after spending some time in Paris with Sara she decided to go to Zurich, to rest and to visit an old friend. The friend was the father-in-law of the first physicist I ever met, Valia Bargmann. Valia was one of Einstein's principal assistants. He had followed him to Princeton and after a while he became professor of mathematical physics there. Sonia, Valia's wife, had been a friend of Yvette's in Germany, and her mother had gone to school with my mother in Russia. Valia was a kind and gentle man. Sonia seemed to me to be a quite cold and unhappy person. I think that she too may have been bored. She had a Ph.D. in chemistry but could not get a proper job in the women-excluding atmosphere of Princeton, except during the war when she worked as a programmer for John von Neumann.

With Mother in Zurich, I decided to try to visit one more cosmic ray laboratory, that at the Jungfraujoch, in the Bernese Oberland. Several of the members of the Birkbeck group were planning to spend the summer there. I called them and they said sure, I could spend the month of August with them. The Jungfraujoch is a high-altitude laboratory used for glacier, weather, and physiological studies, as well as for cosmic ray research. The research laboratory was run by a Swiss foundation in Bern, and they gave me permission to live there and study nuclear disintegrations caused by cosmic rays in photographic emulsions (I had obtained a stack of emulsions in England). What I actually wanted to do was to live at high altitude, and I knew I could learn a great deal of physics from my Birkbeck friends.

The Jungfraujoch is reached by a cog railway from Lauterbrunnen, then to Wengen, and on to Kleine Scheidegg; and then in a breath-taking finale, much of it inside the mountain, one climbs to Jungfraujoch, at 11,300 feet. It is a ride taken by hundreds of tourists daily during the summer months. Many of them turn blue at that altitude from the lack of enough oxygen, and so did I. The underground train station is surrounded by corridors which few tourists see. I was greeted by my friends and guided through one of these to a door marked "Jungfraujoch Hochalpine Forschungsstation—KEEP OUT." The Cerberus of the lab was a man named Wiederkehr who became a legend to those of us who are alumni of the place. He was an elderly Swiss with a very, very bad temper and a surliness to match. It was *his* station and we were interlopers. Once, he became very irritated and carefully removed a door from its hinges, placed it gently on the floor, and stomped on it. Our theory was that the lack of oxygen at the Joch had gotten to him. He was unmarried and virtually never descended to the valley. The Bern foundation, we were told, each year provided a pleasant middle-aged woman to see to his housekeeping needs.

The Joch, in addition to lab space, had individual cell-sized compartments for sleeping, a paneled library, and a lounge cum dining area cum kitchen. All had fantastic views of the Aletsch Glacier meandering down toward the Rhone Valley via Paradeplatz, and the mountain peaks seemed incredibly near. We were generally above the clouds. A secondary lab was located at the Sphinx, a small building some 400 feet higher and connected to the Joch by an elevator. It is there that we placed our equipment and moved hundreds of pounds of graphite blocks to provide suitable shielding. An open platform area next to the Sphinx was used for weather observations: it also allowed a superb view over most of Switzerland. If you stand at Kleine Scheidegg and look toward the Jungfrau, the small building you will see is the Sphinx. The Joch is carved out of the rocks below and faces south.

It was an exhilarating time. There were three physicists from England and I. I liked all three but I developed a crush on one, a wonderfully nice man with a red beard named John Barton. We had magnificent political discussions. John was a socialist, another man was a communist, and I insisted on listening to the Voice of America. Cooking was a little difficult. At that altitude almost everything had to be cooked in a pressure cooker, and everything tasted pretty much the same. We also liked to play poker, but Wiederkehr said absolutely no to poker chips. So in our weekly food order to the valley we included an urgent request for additional packages of noodles, and the motto was: "I'll raise you a noodle and call your bluff." We were allowed one shower every three weeks. Using the toilet was somewhat intimidating. Because snow had to be melted to provide water, a siren was installed to discourage us from using it, which would go off automatically whenever the door to the toilet was closed.

We soon became almost identically dirty, and we

clomped heavily down the corridors, scowling at the tourists like the natives. On the night before I was due to go back to civilization, we bought four bottles of wine from the tourist restaurant, and drank them all. It was our first alcohol in three weeks, and it had the effect we should have expected. It made us blind drunk. I remember crawling to my compartment and, thank god, locking it. John began to pound on the door to let him in. I probably would have had I been able to find the door. The next morning my three friends poured me into the railroad car, and I spent several days in the valley with a devastating hangover. The summer of 1949 was also the last time I saw Lida, though we corresponded for another year or so. Our political differences had become too great. I have often thought of trying to talk to her again, but I never did and she is now dead. Lida was a very important part of my life, and I very much regret that we did not remain friends.

In the fall I returned to Wisconsin. I developed my emulsions and they did indeed show many interesting events, but I needed guidance in analyzing them, and I couldn't get it at Wisconsin. The most exciting group at Wisconsin was the nuclear physics group, and so I decided to try to become a nuclear physicist. The senior professors were Heinz Barschall, Ray Herb, and Hugh Richards. Heinz had worked at Los Alamos during the war and had started a series of path-breaking measurements of neutron cross sections, the experimental foundation of the optical model. Ray had radically modified a particle accelerator devised by Robert Van de Graaff in the thirties: his ever improved electrostatic generators provided the only means to study nuclei with accurate probes for the next thirty years. Hugh was beginning to study the spectroscopy of nuclei, the ways in which they absorb and emit energy.

Heinz's thesis students were generally the best in the department. I felt I could not qualify. Ray's students were primarily interested in accelerator development. I asked Hugh Richards to take me on, and, to my delight, he did. He was then, and he has remained, a wonderfully kind man with rigorous standards. He is an upright man, with strength and common sense. He later became the chairman of the department for many years because the entire faculty trusted him. At first I worked as a technician for an older graduate student, Gerson Goldhaber, and then I began to work as a junior member on some nuclear studies.

Hugh wanted to study nuclear reactions in which neutrons are emitted. Neutrons are uncharged particles. Most detectors of particles use their charge to give a signal. In 1949 the best way to measure the energies of neutrons was to allow them to collide with protons in photographic emulsions—that is, in the coating of special photographic films—and transfer their energies to them. The energies of protons could be measured quite easily, and the energies of the neutrons were then calculated from the results. (This was a technique developed by Hugh Richards at Rice, where he received his Ph.D., and then at Los Alamos, where he worked during the war years.) It worked out as follows: suppose that you want to study the ways in which boron-10 absorbs energy (that is, suppose that you want to determine the quantum states of the type of boron whose nucleus has five protons and five neutrons). You bombard a layer of beryllium-9 (4 protons and 5 neutrons), with hydrogen-2 (one proton and one neutron), accelerated to an energy of several million electron volts in Ray Herb's accelerator. If you observe that neutrons are emitted in the interaction, then boron-10 must have been created. By the law of conservation of energy (you end up with as much energy as you started), you can figure out the quantum states of boron-10 if you can determine the energies of the neutrons.

It was a very straightforward problem. We placed emulsions along the directions of travel of the neutrons. The neutrons penetrated the emulsion but, being uncharged, they did not leave a trace in it. However, when a neutron collided with a proton, the proton took off, and along its trajectory it created a disturbance in the emulsion, similar to that of a contrail when a plane traverses the sky, and we observed a black line when the emulsion was developed. I sat at a microscope and measured the lengths of these lines and the angle each line made with the known direction of the neutrons. I became a superb scanner, able to measure a hundred proton tracks an hour. I was even paid to do it. I once calculated that my salary was a penny per track. It was incredibly boring, but I could do it without thinking and I listened to music at the same time.

The work was boring but the results were not. I was such a fast scanner that the usual problem with emulsion experiments—poor statistics—did not prevent me from getting meaningful results. Hugh was great. He began to give me simple problems of my own. The most important for my self-esteem, which was generally very low, was to try to make a target (a layer) of lithium-6. The lithium found in nature is mainly lithium-7. The lithium which is of use in thermonuclear processes, including weapons, is lithium-6. There are methods for separating the two, but the lithium-6 available for basic research was in the chemical form of lithium-6 sulphate. This meant that it could not be used for precise studies since the signals from the sulfur and the oxygen in the sulphate would overwhelm those from the lithium. At least one other graduate student had tried to solve the problem and failed. I felt desperate but finally I began to think. Could I chemically change the sulphate into something less complicated? Sure, a sulphate can be quite readily changed into a chloride (you mix the compound with barium chloride, and centrifuge out the lithium chloride). (My four years of chemistry were finally of some use!) And what good was that? Not much—the chlorine signals would overwhelm us. Then I decided to check the chemistry library. I wanted to find out everything I could about lithium chloride. And there to my amazement I found that in 1899 a chemist had separated ordinary lithium chloride into lithium and chlorine by electroplating in a bath of pyridine.

I went hunting for pyridine, about which I knew nothing. One of the chemistry professors said, sure, he could give me a flask of it. When I smelled it I wished I were somewhere else: it smelled like the cadavers in a morgue where I had gone on a date with a pathologist. I had thought that this was my first and last encounter with that smell, but I got to know it well because the process worked miraculously well, and I had clean, elemental lithium-6 targets.

Pyridine did have a few minor problems: I found that in the nineteenth century it had been used as an asthma remedy (that smell could really clean one's bronchi) until it was learned it also led to heart failure. Also, pyridine catches fire easily, as I found out to my chagrin once during a colloquium I had missed because I wanted to get on with my experiment. I put out the fire myself but instead of giving me an award for heroism Hugh Richards asked me why I had not been at the colloquium.

There was also the time when Ray Herb, followed by a retinue, painted the gas pipes in the basement with a soap solution to locate a gas leak (a leak would make bubbles). For a while I did not understand what was happening, since I had lost my sense of smell. I never did tell him that it was the pyridine seeping through the door to my laboratory.

Hugh was very pleased with my success and made me publish the method, but very quickly we got reports that my results could not be repeated by other groups. There is hardly anything more awful to a scientist than to have published unrepeatable results. It turned out that in order for the method to succeed the pyridine had to be exceedingly pure, and by incredible luck I had happened to have received some of the world's purest pyridine from one of its leading experts.

While my research was going well, my academic work was poor. Once again I failed a course, this time in quantum mechanics. The course grade depended entirely on the final examination, which I took when I had a bad case of flu; in any case, I did not know the material well enough. I could not understand any of the questions on the exam, much less answer them. I think I had amnesia, but I did not try to excuse myself, and I quite properly failed the course. I did learn enough quantum mechanics to pass the Ph.D. qualifying exam, which I delayed as long as possible, to the beginning of my fourth year at Wisconsin. By then it was probably a guaranteed "pass" because under Hugh Richards's direction I had done a couple of experiments on boron-10 which demonstrated that some very beautiful work done by a Caltech group headed by T. Lauritsen and W. A. Fowler was wrong. The sequence of states I then determined is still correct, and there was considerable theoretical interest in our results at the time. I say "our" results advisedly since Hugh had more to do with their interpretation than I did. But he insisted, over my strong objections, in not putting his name on the resulting papers. He had then, and continued to have throughout his career, the notion that the graduate students should get all the credit. He is a very unusual professor!

I think it is clear from this account that the interaction between professors and graduate students was unusually close at Wisconsin. In part this was because we were much the same age at the time, because we were the first postwar generation. Also, they were fine people who loved physics and were socially relaxed. We were all on a first-name basis, and most of our social life was integrated. We had parties and picnics together. For a time I had a small apartment in Sterling Court with a tiny Frigidaire in the bedroom and a universal sink in the bathroom. There were cockroaches and the electrical wiring was such that when I put on my window air conditioner, Heinz's lights (he lived in an apartment upstairs) went out and he would come downstairs to berate me. Nevertheless, a lot of good parties took place there. Sterling Court is no more: it has been replaced by an antiseptic university building.

We also had a yearly Christmas colloquium in which we sang songs and played skits skewering the faculty and ourselves. Ray Benenson's song (played to the tune of "On Wisconsin") about what we would do to Herb's machine when we were through with our theses was a real winner. I may unwittingly, in fact, have precipitated the collapse of the accelerator. I needed one more day of running time but the previous user had left it with a leak in its vacuum system (this was part of the tradition: you did your own work; the next guy/gal was on his/her own). I finally figured out where the leak was probably located, but I could not fix it by the usual methods. So I dissolved a piece of my plastic raincoat in acetone, dripped it onto the area of the leak, and tied some knitting yarn into the goo. It worked for me. But when I left, that part of the machine collapsed.

We were encouraged to go to meetings of the American Physical Society and give talks, but no special funds were available. So we chipped in, got into a car, drove to the meetings, stayed at cheap motels, and generally had a glorious time. Most of the people I knew at that time have stayed on in physics, and some of them have done distinguished work. Among them are Bob Adair of Yale, Gerson Goldhaber of Berkeley (who also became an excellent particle physicist), and Gene Amdahl, who founded his own computer company.

I loved nuclear physics but the question was whether I could continue to be active in it. I knew that I wanted to teach. I had taught some sections at Wisconsin, and there was my Navy Pier experience. I did not know how one went about getting a job in academia. I thought that I could graduate in the spring of 1952, so in fall 1951 I asked Hugh what I should do. He suggested that I return to Europe or, failing that, that I apply to a women's college. I was tremendously depressed by his comments. I did not realize until later that what he was giving me was the best possible advice at the time to a woman. Going to Europe was out: I was an American and had no taste for Europe. It was true, however, that women scientists were better accepted there. Most institutions in the United States, other than women's colleges, would not accept women as faculty. So I applied to Bryn Mawr and to Smith College. Bryn Mawr's offer was for a research fellowship (though they had no ongoing research);

they added that if I wished, I could teach at no salary. Smith offered me a one-year appointment to replace someone who was on leave; I accepted this offer for the fall of 1952.

Then I had an idea which determined my future. I wrote to Tom Lauritsen at Caltech and asked him if he might be willing to hire me for the summer to work on a rewrite of an article he had published two years earlier with some of his colleagues, summarizing all the work done on the nuclear spectroscopy of the so-called light nuclei (nuclei lighter than magnesium). He knew of me through the work I had done in which I had, sharply and untactfully, referred to the incorrect Caltech results on boron-10. Tom was quite a man. He hired me. We worked together until his death twenty-one years later, and I then continued the work without him for another seventeen years.

So I had a job for the summer and a job for the following academic year, but I still had to complete my Ph.D. thesis and pass the thesis exam. The problem was not with my thesis. I had published enough so that Hugh said I could just collect the papers I had written and write some suitable covering material and appendixes. The problem was that I had, contrary to Wisconsin tradition, not really built any equipment. I had just used it. In fact, I had even used it untraditionally. When I operated Ray's machine for long runs there wasn't too much to do and I was bored. So I knitted—until the day Ray sent a notice to *all* members of the nuclear group stating that people who used the accelerator were, please, not to knit at the same time. I met Ray in the hall about a month before my Ph.D. exam was to occur, and he said, "Gee, Fay, wouldn't it be nice if you built a one-quarter MeV generator before you leave." I didn't know if he was kidding. It would have taken me years to build one. I disappeared into my apartment and typed furiously. I avoided Ray. I avoided all but Hugh. I came in for my exam and passed it. Ray was there but didn't comment on my lack of equipment experience. Much later I found out that Ray did indeed want me to build that machine but, being a very nice guy, he decided not to press the issue. I left for Caltech without waiting for graduation.

During my Wisconsin years, Misha had begun to do extremely well. He set up a company which, in the 1950s, acquired a factory about 30 miles from New York and began to build motor-generator sets. The company was held by him and he placed me on its Board of Directors, which involved very little work on my part and amused me. Mother was bored once again but could no longer have a large effect on my life. So she decided to work on Yvette's. She prodded my sister to finish her college work at Columbia. Iva is very bright indeed. Though she was then in her forties she not only finished her undergraduate work but received her Ph.D. in a total time of about five years.

3

100
Memorial Drive
1952–1957················

I DID NOT HAVE A CAR. In fact, I had barely learned to drive. I went to Caltech via buses and trains, seeing a great deal of America along the way. I stopped in Denver and spent a day at another of the cosmic ray observatories, at Mount Evans. I was traveling light and without reservations. Back in Denver I found that a Republican convention had booked up space in all reasonable hotels, so I stayed for one night in a *hotel-de-passe*. My room reeked of cheap perfume and cigars, the sink and toilet were in questionable condition, and, in addition to locking the door, I placed a chairback against the doorknob. I put my mind in neutral and fell asleep, disregarding the noises I heard through the thin wall from the adjoining room. The next morning I quickly took a bus south and then a train to Lamy, in New Mexico.

Heinz Barschall and several of his graduate students were spending the summer working at Los Alamos, and Heinz had invited me to drop by on the way to Caltech. To me Los Alamos was a legend, a place high up in the sky where a group of extraordinary physicists had succeeded in changing history. I was conscious of the problems that nuclear weapons would bring, but I had no qualms about the destruction they had already caused. Japan had attacked America. Japan had slaughtered the Chinese. Japan was an ally of the Nazis. The bombs prevented the killing of half a million American troops. I was proud of what had been accomplished at Los Alamos.

Lamy is the rail stop for Sante Fe. In 1952 Santa Fe

was a quiet and very attractive town. Los Alamos was 35 miles away on a mesa reached by what was then a rather rudimentary road. It was still a closed, fenced-in town, with a guard tower at its entrance. Heinz had arranged for a pass, but I had to be escorted whenever I visited a laboratory building, even though none of the work I was interested in was classified. People were extremely nice, and I was asked whether I might like to come and work at Los Alamos in 1953, after my one-year appointment at Smith College. I was grateful but I wanted to teach and not just do research. Heinz took me to visit the San Ildefonso pueblo near the bed of the Rio Grande and introduced me to PopoviDa who had been one of his technicians during the war. PopoviDa's mother, Maria Martinez, and he were creating beautiful black pottery, which later became extremely well known. I bought a dish for ten dollars, a great deal of money for me. I still have it, and I have became entranced by such pottery, by Hopi Kachinas, and by Two-Grey Hills rugs. It became a lifelong interest. We said good-bye at a dinner organized by Heinz for me and for his students at a private hacienda in the valley called the Swan Lake Ranch. It had a huge fireplace with burning logs. The evenings and nights were very cool at Los Alamos because of its altitude. I have returned many times, and, for eighteen years, I was a visiting staff member. I still love the place. I think of it as Shangri-La, despite the fact that I have terrible headaches and difficulty in breathing during the first week. Perhaps it was so in Shangri-La as well.

Heroes are quite out of style today. But I have been very lucky, I have had three.

The first was Misha, the second Tom Lauritsen, and the third my husband, Wally. I met Tommy shortly after checking into the Athenaeum, the Caltech faculty club, where he had made arrangements for a room for me. He was then thirty-seven years old, an associate professor at Caltech. He died in 1973 and I miss him still. Tom cannot be described in a few words. Willy Fowler and I called him "a man of integrity, responsibility, decency and gentle wit."[4] This was true, and he was also intelligent, kind, courteous, and possessed of a quite ferocious sense of humor. He would have been perfect had he only been slightly less kind. I adored him immediately and became his apprentice. He taught me much of the nuclear physics I know, and he developed my taste for good physics. Tommy had gotten me the appointment at Caltech by neglecting to tell its troglodyte dean that I am a woman. (This was made possible by my name which, in current jargon, is not gender-specific.) I integrated Caltech. One of the fun consequences was that there were no rules governing how women should be dressed when lunching at the Athenaeum. Men were required to wear jackets and ties in the heat of that unairconditioned building. I sauntered in, in sandals and beach dresses.

Caltech was an incredible place. Kellogg Laboratory was the premier lab in nuclear physics. The senior professors were Charlie Lauritsen, Willy Fowler, and Tom Lauritsen, shortly to be joined by Tom Tombrello. Bob Christy was the resident theorist. Charlie, Tom's father, had created the Kellogg Lab from scratch in the thirties. He had come to the United States as an immigrant from Denmark with his wife, Sigrid, who became a physician, and with Tom. After working as a high-level technician, he was attracted to Caltech by R. A. Millikan and became a superb nuclear physicist. Charlie had been the Ph.D. supervisor for both Willy and his son, Tom. Willy Fowler went on to win a Nobel Prize in 1983 for his work on nucleosynthesis (the formation of elements in the universe). He is a delightful, feisty character, with very strong views about practically everything, and an uninhibited sense of humor. For instance, shortly after I married I introduced my husband to him at a meeting. He said "Hi" and then peered into my decolleté and said, "My, those are two big beautiful ones." I thought it a friendly and quite hilarious and unsexist remark, because I also knew how much he had helped the careers of the women scientists who had worked with him. He enjoyed excellence and, in that connection, gender was irrelevant.

Tom Lauritsen worshipped his father. Charlie was an extremely clever hardware man and he had developed a lot of clout in determining national science policy. During the war, Charlie, Willy, and Tom had set up labs in Pasadena and nearby which designed and built most of the rockets used by the U.S. Navy during World War II. Charlie always seemed to be taking the red-eye special to Washington. He was a friend and supporter of J. Robert Oppenheimer. (A bit later, during the Oppenheimer affair, the media concocted a scientific mafia nicknamed ZORC, for Zacharias, Oppenheimer, Rabi, and Charlie Lauritsen.)

Mother came to visit me at the Athenaeum. She was always impeccably dressed. I rebelled by usually dressing frumpily and, in my adult years, I have worn stockings only once. So Mother decided to apply a little pressure. She asked Tommy to tell me to wear stockings. That wonderful man said to her, "Mrs. Ajzenberg, I have been trying for years to get my wife to wear *shoes!*"

Mother told me that Willy had asked her whether I was in love with Tommy. I was horrified at the thought since Margie, Tommy's wife, had become my best woman friend (as she still is). I tried to imagine Tommy romantically, perhaps even with male organs whose shape I was beginning to deduce, but I couldn't. I had been so thoroughly brainwashed as a child that I have had strong sexual desires only for the person who became my husband; although to be honest, there was one other man I found physically attractive. He made a gentle pass at me, but because I was married to Wally I didn't respond.

Margie was Tom's second wife. His first wife, a Dane, had died shortly after giving birth to a son, Eric. Tommy then met Margie, who during the war had been a sergeant in the air force, stationed in India. They had two children. All five lived in a sprawling old house in Altadena. Most evenings Tommy would call her and say, "I'll bring X people to dinner." The X included me, graduate students, and visitors. Margie never seemed to mind. She appeared unflappable. She added another pound or two of spaghetti to the evening cauldron, picked some more avocados from the tree outside, and got out the gin and the beer. I remember the summer as consisting of informal evening parties at which there was a lot of drinking and talking, with occasional singing of grave Danish songs by Tom. In the morning I would stagger into my office with a hangover and find my friends clean-cut, bushy-tailed, and already hard at work.

The post-World War II period was an extremely exciting time in physics in general and at Caltech in particular. Physicists were no longer working on weapons. Instead, they were creating new fields and bringing fresh ideas into old ones. Advances in techniques that were developed for applied uses opened new vistas. Contacts with scientists from other countries were resumed. Government funding of basic research, which had been negligible before the war, became easily available: the

admirals, the generals, and the political leaders had become convinced that science was useful to the country.
Charlie Lauritsen's influence led to the Office of Naval
Research. Bob Bacher, another professor at Caltech, was
a commissioner of the Atomic Energy Commission, which
later became the Department of Energy. Other physicists
started the National Science Foundation. And at Caltech
new fields were being created. Willy Fowler started the
field of nucleosynthesis. Dick Feynman and Murray Gell-
Mann were doing superb work in the new field of particle physics, and they too, like Willy Fowler, eventually
became Nobelists. Astrophysics and geophysics at Caltech
were world-class. To a large extent, Kellogg's people were
central to the intellectual vigor of Caltech. The halls of
Kellogg throbbed with excitement.

My interest in nuclear physics had focused on the light
nuclei, which I shall arbitrarily define as nuclei with not
more than twenty neutrons and protons. Because these
nuclei are made up of relatively few particles, their quantum structure, which arises from the ways in which they
can absorb and emit energy, is relatively simple. It was
becoming clear that theoretical models of nuclear interactions could be tested by comparing their predictions
with the experimental data on the quantum states of the
light nuclei. An understanding of these nuclei was also
becoming essential to predictions of nucleosynthesis in
stars, and to resolving the most fundamental questions
about cosmology and symmetries in nature. Applied uses
ranged from "dating" techniques (using the isotopes
carbon-14, and later beryllium-10), to nuclear medicine,
and to nuclear fusion (in fusion reactors as well as in
weapons).

My work with Tom consisted in analyzing all the papers published on the quantum states, the "energy lev-

els" of the light nuclei, digesting the information, and writing a manuscript in which we presented our views of the best information about the nuclei. It required the reading of every paper published in the field, judging the validity of the work, seeing how it fitted in with the earlier data, and stating our case. Our first scientific review paper, based on that summer's work, in 1952, was eighty pages long. It was published in the *Reviews of Modern Physics*. Our later papers were an order of magnitude longer. By now I have published some twenty-six of these review papers, totaling some five thousand pages, principally in the journal *Nuclear Physics*.

In retrospect, I find it amazing that we finished the first paper so quickly: this was our first collaboration and I was relatively ignorant. Tommy claimed at the time that I beat him regularly to complete the paper. At the slightest provocation, he would pull up his shirt and show the invisible lashes on his back to everyone present.

We evolved a procedure which consisted of my writing a first draft of the information on a particular nucleus, and presenting it to him with copies of the relevant papers. A few days later I would receive a second draft, which bore very little resemblance to the first. Then I painfully tried to figure out how I had screwed up, and accepted the second draft with few modifications. I have his early scrawled comments still: the letters HST (horseshit) are prominent. Tommy not only said that I beat him but he kept saying that I stole his pencils. I finally presented him with a box of one thousand pencils and had some peace for a while. I drew poster-sized diagrams of the energy levels of the light nuclei, which were then transformed into beautiful ink drawings by Barbara Zimmerman at Caltech. These diagrams became the figures of our review articles. The poster-sized

copies lined the main corridor of the Kellogg Lab, and
they were updated as we received new information from
our friends all over the world. Occasionally there were
scrawled comments about the validity of various experiments, as well as pungent asides about the antecedents
of certain scientists.

I felt an almost sensuous delight in looking at the drawings as they evolved month by month: the lines moved
up and down and the numbers changed as better data
became available. All of a sudden one could look at one
of these pictures and say, "Yes; this is *right*. This is beautiful. This region of this nucleus is *understood*." It is perhaps similar to the way in which an artist can look at a
picture and say, "This is real. This cannot be a fake."
Recently attempts have been made to codify the evaluations of nuclear data, that is, to set up rules which state
under what conditions a particular nuclear property (for
instance the angular momentum of a quantum state) is
labeled as "known." This means that the evaluator of the
data can rely on the rules, and do the job more quickly.
While such attempts are rational, I do not know how an
algorithm can be devised which can take into account
such intangibles as "Do the new results 'look' right when
compared to what we knew?" and "What's the track
record of the scientist who obtained these data?"

At Caltech I worked on the energy level reviews in an
office without windows that had been a storeroom at an
earlier time. The summer of 1952 was marked by a series
of strong earthquakes in Southern California and hundreds of aftershocks. In the closed box of my office I
seldom knew whether I felt seasick because the building
was moving, my hangover from the previous evening's
party was worse, or I was developing a bad case of claus-
trophobia. At Wisconsin I also worked in windowless

ex-storerooms (as soon as I cleaned one up and made it comfortable, I was moved to another), but there at least the rooms didn't move. On the other hand, my office at Kellogg was centrally located and Dick Feynman dropped in when he needed an audience. He was mesmerizing: he once talked to me of magneto-hydrodynamics for four hours, and it was exciting and seemed crystal clear. Of course when he left I realized I hadn't really understood anything. Listening to Dick then, and on a number of occasions afterwards, always gave me a high. He was a superb physicist and a complicated but modest and warm man. He was really a genius, and he was a profoundly human person.[5]

I also worked with Tommy during the summer of 1954 at Caltech, and we finished another paper. Then, until his death in 1973, we worked at long distance. We had developed a symbiotic relationship, and cryptic remarks, some more subtle than HST, were all that we needed to understand each other. In 1954 the records of the Oppenheimer hearings had become available: Caltech, where J. R. Oppenheimer had once taught, was a hotbed of pro-Oppenheimer sentiments. In fact, virtually all physicists were united in opposition to the tenor of the hearings and virulently opposed Edward Teller, whom I later came to like and admire very much. The only anti-Semitic remark I ever heard at Caltech occurred during one party when the very drunken wife of one of the scientists said, "It is all the fault of that damn Jew, Strauss" (then the chairman of the AEC), having managed to forget that Oppenheimer was also a Jew.

But life at Caltech was not all physics and science politics. It was also about friendship and about playing tricks on one's friends. Practical jokes were a Caltech hallmark. They were a way of letting off steam in an otherwise

nerdy, cloistered, and competitive atmosphere. Jokes relieved tensions and strengthened the affection that bound
faculty and students. One of the better jokes was cooked
up by the "Dirty Five," a group of graduate students
living in a house near campus, and me. (Once one was a
part of Caltech, there were no closed doors between students and faculty; and my gender, if it had any relevance
at all, made it easier to be welcomed by both.) Willy
Fowler was going overseas with his wife and two daughters. They were going east by train from the Pasadena
station. Willy had a passion for trains and took trains
whenever possible all over the world. He also liked champagne. So we bought a bottle of good champagne, drank
it, filled the empty bottle with water, and carefully replaced the cork and the foil. We went to the station, saw
the Fowlers off, and at the last moment handed Willy
the bottle. However, we deliberately dropped it as we
handed it to him. His face as he witnessed the destruction of a bottle of French champagne was a study in
horror. The "Dirty Five" and I had another triumph.

The people who shaped me as a physicist were my
teachers and colleagues at Wisconsin, Caltech, and then
at MIT, but Caltech is the institution to which I feel that I
owe my scientific life. The summer of 1952 was my
epiphany. I realized then that it was possible for me to
be a good nuclear physicist. The experimental work I
had done at Wisconsin had become well known. From
Tom Lauritsen I had learned to read and to evaluate the
work of others. I corresponded with scores of other
nuclear physicists in this country and abroad. I felt pride
that my work could be useful to them. I could think of
hundreds of scientific questions to ask, and I had even
begun to understand that the likelihood that I would get
answers was as good as my colleagues'. However, the

most important reason for the confidence I felt was that Tom Lauritsen and Willy Fowler respected me as a scientist. Once Willy even crowned me an honorary alumna of Caltech, with a beer bottle.

In the fall of 1952 I visited my parents in New York on the way to my new jobs in Massachusetts. Misha took me to the window of the living room, pointed to the parking lot below, and asked "Which is the most beautiful car there?" I pointed to the newest car. He beamed and told me it was mine. It was a stripped-down Ford Falcon, and I loved it from the first. Misha was doing well again. He took us to a good restaurant and said, in Russian, the code phrase that he used whenever he was flush: "The management does not stop at any expense," to which we responded, "Fifty girls, fifty." It was alleged to be a common advertisement for burlesque houses in Russia, though I find it difficult to imagine such premises there now. The phrase meant that we could order anything on the menu, regardless of cost.

I drove my car to Boston. Two of my classmates from Wisconsin held post-doc positions at MIT, in William Buechner's Van de Graaff Laboratory, working on nuclear reactions. (Ray Herb's machine was called an electrostatic generator at Wisconsin and a Van de Graaff at MIT, where VDG had done his work in the thirties.) My two friends suggested that Bill appoint me a Visiting Fellow and that he pay me enough to have apartments both at Smith College, in Northampton, and in Boston. As I recall, this amounted to a little over $100 a month, and I greatly welcomed this opportunity to continue to do research. I got a small apartment on Storrow Drive in Boston, and I also found a tiny apartment in a rooming house near the Smith campus.

Most weeks I got up at 5 A.M. on Monday morning,

drove 90 miles to Northampton, started to teach at 8
A.M., and continued to teach until Wednesday afternoon,
when I drove back to Boston, and to MIT. I learned to
drive that year. There was no thruway and the roads
were poor—and slippery during the winter. My life became very compartmentalized. When I was at Smith, I
taught, graded, and prepared my lectures. I had the appropriate clothes and supplies in my Northampton apartment. At MIT I listened to colloquia and seminars, did
research and analyzed data. Both my apartments were
self-sufficient. I just somehow had to manage to get myself back and forth.

I liked the faculty and the students at Smith but I had
very little interaction with them. I knew that I would
leave after two semesters, and that my opportunity to
remain in physics depended on the research I did at
MIT. Smith reenforced my desire to teach, and it also
gave me a very strong signal: if I wanted to continue to
be a physicist, it could not be from the base of a women's
college. Colleges at the time did not value the research
activities of their faculty. The usual teaching loads were
two to three times those of faculty at universities. And
the women students at Smith, while outstandingly bright,
were not career-motivated. I felt very frustrated that I
could not interest them in becoming working physicists.

At MIT, on the other hand, life was great. I was having a fine time at Bill Buechner's lab, and I also worked
briefly at the electron accelerator headed by Lou Osborne
and Peter Demos. The intellectual and scientific ferment
of Cambridge was a delight, and I wanted to remain
there. I called the chairman of the Department of Physics
at Harvard, Ken Bainbridge, and asked him if an in-
structorship might become available. He was very pleas-

ant but said no. I was very naive. The physics department at Harvard has hired a few junior women faculty following the advent of federal antidiscrimination laws in the 1970s. They did not give tenure to a woman until 1992. Of course, their argument would be that none had qualified before, but there are plenty of male faculty members at Harvard, or anywhere else, who are second-rate. I will believe that discrimination against women has stopped when I observe that second-rate women are given tenure. My preference is not to hire second-rate people of either gender. A giant step forward will occur when excellent women have no more difficulties than a man would in obtaining tenured faculty positions in physics departments at research universities.

A position as assistant professor was available at Boston University. I applied for it, was interviewed by the chairman, Dow Smith, and hired. Dow is a terrific person and we hit it off right away. He was pleased by my desire to set up a small nuclear group at BU and continue to work at MIT as a visiting scientist. He offered me a respectable salary and I accepted it. When the contract came back signed by the dean it was for 15 percent less. When I queried Dow, he told me that the dean had said that as a woman I needed (and presumably would accept) less money than a man. Misha had driven into me the notion that one should only work for fair pay. He felt that people do not respect those who are willing to be underpaid. And Peter Demos and Lou Osborne had told me that if BU did not work out I could stay on as a post-doc at their lab. So I was able to ask Dow to tell the dean what to do with the contract. The dean changed his mind, giving me the opportunity to enjoy working with Dow for four years. Many years later Dow told me that he really hadn't wanted to consider a woman, that my interview with him was pro-forma, but that because we clicked he decided to hire me.

My life changed completely. I began to think that I would stay at Boston University permanently and that it was all right to have long-range objectives. I found a very pleasant apartment at 100 Memorial Drive in Cambridge, adjoining MIT and facing the Charles River. Rents there were reasonable and the arrangement of the building was a delight. There were corridors every three floors, with many doors. Some opened to apartments on the same floor, others to small individual staircases going up to the floor above or down to the floor below. I had one of the "up" apartments with a living room and kitchen overlooking the Charles to the east, with a large balcony. The bedroom faced the Charles to the west. There was no air conditioning, but the cross ventilation was marvelous. The apartment was not furnished and Misha began to visit and to select and buy furniture for me. Sometimes he just chose it by himself and had it shipped. Sometimes he took me on shopping expeditions in Boston and we would decide what to buy together. He was adamant that the quality of the pieces must be very high, and most of the furniture, rugs, and art he bought are with me still, and I still enjoy them. I look with amusement, however, at a splendid vanity table he sent me. He wanted me to be beautiful and well groomed, as well as a scientist. He kept suggesting Madame Kollontai as a model. She had been a very prominent early Bolshevik and a very beautiful woman. Although my father loathed communism, he was somehow entranced by Kollontai. He admired her combination of beauty, toughness, and brains, and presented her as a role model for me. I recently read a biography of Madame Kollontai, and I think that Misha really did not know a great deal about her: she was quite a ruthless person, and she was very promiscuous. He disapproved of such characteristics in either men or women. The vanity table's mirror has been in the "down" position for at least twenty years.

Misha worried about me. I had become almost totally exhausted physically after commuting to Smith for a year. In addition, I had severe ragweed allergy problems: as soon as the season started in August I would begin to sneeze uncontrollably and I had asthma attacks well into late fall. I tried to get desensitized by taking pollen shots all year long, and I did so for seven years, but all that this meant was that I had swollen arms every week and I was never ready for the ragweed season. Medicines did not relieve my symptoms and they made me feel as if I had a perpetual head cold. In later years I tried all sorts of creative ways to disappear during the season, for instance by taking the ship *France* to Europe and right back again, but this was not always practical. Then my immune system was totally changed by chemotherapy at the age of fifty-seven. My allergy to ragweed disappeared.

But in the summer of 1953, I was exhausted and the ragweed season was coming. Misha took over. He picked me up and we drove north to Maine, to Nova Scotia, and to Montreal. I loved being with him but I was still sneezing and he had to be cautious about smoking his beloved cigars because it only made things worse. One night, to our joint embarrassment, we could only get a double room, with twin beds. I sneaked into mine while he waited outside and, after closing the lights, he dove into his. In Montreal my asthma got steadily worse and he finally drove me home. I was very touched by his love, by his taking the time to be with me, and by his so obviously enjoying it.

Misha was also concerned because I was quite depressed. The work was going well but I had no one with whom to share my life. He decided, when we returned to Cambridge, that I needed to take care of something and presented me with a bird of dubious lineage, whom he named "Proton," and a large cage. This was not quite what I needed. Proton was a surly beast and demanded constant attention. I had a fastidious distaste about cleaning up his cage and the room when he flew around it. I had visions of opening the door to the balcony and seeing if he could survive in the wild. Fortunately Proton took ill and died, after I spent a very difficult weekend chasing around Cambridge and Boston for antibiotics appropriate for bird diseases (this was only out of a sense of duty toward Misha). Misha visited me often during my Cambridge years. He came to talk, and he even had his medical check-ups at Peter Bent Brigham Hospital. He had polycythemia, and the treatment at the time was to bleed the patient at intervals.

It was not surprising that Mother, bored again, became very jealous of the relationship which my father and I had built up. She screamed that I pretended to love Misha only so that he would give me presents—that I loved him for his money. Intellectually I understood what she was doing and why, but I was so tired and depressed that I worried whether she might be even partially right. It inhibited my relations with Misha, though I told him how much he meant to me a few months before his death in 1962. I kept his name after I married, which was quite unusual at the time. I have established a scholarship in his name. And I hope that the work I have done through the years, bearing his name, has been honorable and useful and speaks to my love for him.

Boston University was in the process of changing from a quite parochial institution to a national university. It achieved that status twenty years later under the presidency of John Silber. Its campus was filled by a complex of buildings dominated by two towers called by its inhabitants "the twin penii." The bureaucracy was more appropriate to a high school than to a university, but Dow Smith was a genius at creating an island of civility and scholarship within the physics department. He also directed a lab doing optics research for the air force, across Commonwealth Avenue. My recollection is that it later detached itself from BU and became part of the industrial company, ITEK. My faculty colleagues were good physicists and quite companionable. Some of the undergraduates were very able, and several graduate students worked with me. I had received my first research grant from the Air Force Office of Scientific Research, primarily due to Tom Lauritsen and Willy Fowler's help. My first grant was for $10,000 and it bountifully filled all my needs. My last grant, in 1990, was about ten times larger but I constantly had to scurry to make ends meet.

There are essentially two varieties of physicists who do basic research. In each subfield of physics there are theorists who do calculations based on available experimental data or who predict new phenomena; and there are experimentalists. The latter come in a number of "flavors": in nuclear (and in particle) physics, there are people who are experts in designing and building particle accelerators. Other physicists are interested in creating and operating equipment that will detect particles. Some people are computer experts, and others prefer to analyze the data and deduce their significance. And there are honcho-type people who like directing such groupings of diverse scientists. There is, of course, a certain amount of slippage between the various flavors of experimental physicists: it is possible, although unlikely, to be expert in more than one area.

Most professional physicists can readily see which problems should, and can, be solved at a particular time. To resolve scientific questions involves thinking, of course; but equally, one must have access to people with the needed expertise, to the right equipment, and to adequate financial support. In 1954 I knew what I wanted to do as an experimental nuclear physicist: I wanted to continue to earn my "keep" by continuing to work with Tom Lauritsen on reviews of the international work on the energy levels of the light nuclei. This would be useful to my field. And I wanted to have fun, and that meant being involved myself in determining the structure of the light nuclei.

When I reviewed the evidence on a nucleus, I would suddenly become aware that some of the evidence was poor; or that there was an important lacuna. I would then think of a way in which I might obtain the necessary information. I never had an accelerator of my own, but I had friends at most accelerators in the United States and in Western Europe.

For instance, in the late 1950s, when I was at Haverford College, a great deal of interest had been generated about the second excited state of the nucleus carbon-12. That state, at an excitation energy of 7.66 MeV, is now known to play a key role in energy generation and in element synthesis in red giant stars, through the fusion of three alpha particles into carbon. The questions at the time were: (1) Can this state be formed by the fusion of three alpha particles? and (2) Once formed, will this state decay into a stable kind of carbon? One of the steps in knowing how to solve this problem was to determine how often the state was formed in a particular reaction. I called a friend at Oak Ridge National Laboratory, Paul Stelson, and we ran the Van de Graaff there for a day, irradiating beryllium foils with alpha particles, and exposing photographic emulsions to the neutrons emitted in the interactions. I returned to Haverford College with the emulsions. In a week or so, I had scanned them, and we published a useful paper.

Basically this was what I enjoyed doing: seeing that there was a problem, knowing where and how it could be solved, and participating in its solution. What I could contribute best was my expertise in knowing what was already known, having a well-developed taste for good physics (and for good physicists), and being totally aware of what technical facilities were available, anywhere.

In my four years at Boston University from 1953 to 1957, my students and I published five research papers and gave a number of talks; and I wrote three review papers with Tommy. The research papers were done using Bill Buechner's particle accelerator at MIT, and the cyclotron at Princeton. I needed higher particle energies than were available at MIT for some of my experiments, and I had friends at the Princeton cyclotron. They warmly invited me to come and expose my photographic plates. The analysis would be done at Boston University. It worked out very well except that I had to sneak around the physics department at Princeton. I had been told that Allen Shenstone, the chairman of the department, had a rule: no women in the building. I worked mainly at night; in the daytime I had to be cautious. Princeton now has a

couple of junior faculty women in physics, but it has yet
to give tenure to a woman. My remarks about Harvard
also apply here.

I very much enjoyed working with my Boston University students. I taught a course in nuclear physics, and
when we came to the development of nuclear reactors, I
showed them the famous bottle of chianti that had been
signed by the physicists who were present when the first
reactor, at Chicago, had gone critical. It was loaned to
me by Al Wattenberg, then at MIT. One of my three
graduate students was Gloria Lubkin, who became one
of the world's best known and most respected science
writers. She is editor of *Physics Today*, and one of my
closest and beloved friends. We went together to colloquia
and seminars at MIT and at Harvard, and we visited
other laboratories. On one occasion we had a very interesting visit to Brookhaven, and we stopped to talk with
Gregory Breit on the way, at Yale. He was not inherently
comfortable with young people but that day he was wonderfully helpful and warm. I had gotten to know him
through Eugene Wigner. Eugene used to come to Wisconsin during the summer months while I was a graduate student there. The first time I was introduced to him
I was so awed that I curtsied. He was Professor Wigner
to me and to the other students until the summer when I
volunteered to type his notes, which became known as
"The Wignerian Lectures on the Shell Model." (Wigner
received the Nobel Prize in 1963.) He was the soul of
courtesy and old-fashioned politeness. It was impossible
to get him to walk through a door first. His mind was
exceptionally sharp, and we learned to watch for an
apologetic humming, as his right hand fluttered over his
mouth. He would say to a guest lecturer, "Excuse me. I
am so stupid but I do not understand what you did in

line 5." The wrong response was to explain patiently what one had done in line 5. The right response was to realize that one had done something totally stupid.

Eugene, a Hungarian, had fled Europe during the thirties. It was he who, together with Leo Szilard and Edward Teller, got Albert Einstein to write to President Roosevelt to start the Manhattan Project. He remained an important, and very conservative, adviser to the government. He was quite a modest person. His second wife, Mary, whom I liked very much, was a plain-spoken, sensible and very direct person. It would be hard to conceive of a person more different from the elliptical and subtle Eugene. In the 1960s I finally began to call him by his first name. While I was a graduate student at Wisconsin it would not have worked. After one long typing session in which I had asked him many questions because I couldn't understand his writing, he said to me, "Miss Ajzenberg, I mean Fay, I mean Miss Ajzenberg— couldn't you call me something other than Professor Wigner?" Then he paused, thought for a couple of minutes about the matter, and could not come up with an appropriate alternative.[6]

Victor Weisskopf was the towering figure among the physicists at MIT, and in Cambridge in general.[7] Viki was born in Austria. He had worked at the Bohr Institute in the thirties and married a very nice and classy Dane named Ellen. They emigrated to the States during the late thirties and were at Los Alamos during the war. Viki is a very cultured and civilized man, with a great deal of flair and charisma. His back-of-the-envelope calculations are famous, and he is an extremely good speaker. He is a liberal, interested in fostering international cooperation between scientists. I like him very much. He later became the director of CERN (the

European Center for Nuclear and Particle Physics, in Geneva) and was instrumental in making it a frontier facility where people from many countries, including the USSR and China, worked together. Viki and Ellen were extremely hospitable, and there were often parties at their house, to which even young scientists like me were warmly welcomed. In this way we met many world-class physicists.

Weisskopf's tradition of including young scientists was also followed by David Frisch, another professor at MIT: he only invited junior people to meet Enrico Fermi, who came to Cambridge shortly before his death. Fermi was a revered figure, and it was wonderful to be able to hear him in a small group. The only time I skipped a Weisskopf party to which I was invited was when Werner Heisenberg came to town. Heisenberg was a great physicist but he had been deeply involved with the Nazis. Even though the war had been over for some time, I thought it inappropriate for Jews, in particular, to welcome him. I explained my reasons to Viki and did not go. As far as I know only one other person joined me— Herman Feshbach. Herman, who became a close friend, had a great deal of guts, since he was then a junior member of the MIT faculty. He later became chairman of the physics department at MIT and one of the most influential physicists in this country. Herman was indirectly responsible for my realizing early on that I was incapable of serving in stuffy, ceremonial roles. At a banquet of the American Physical Society in 1973, I was sitting next to Herman on the dais. When the president of the APS began to eulogize Feshbach, prior to bestowing an award on him, he referred to Herman as one of the world's premier experimental nuclear physicists. (Herman is, of course, a leading theorist whose work, on the optical model, was based on the experimental data obtained by Heinz Barschall and his group.) I found this hilarious and began to puff with laughter, while Herman, who was trying to display an appropriately serious mien, muttered from the side of his mouth, "Shut up. *Shut up!*"

Viki was very much involved with the Federation of American Scientists, an organization which lobbies on scientific/political matters. He thought that there should be an active chapter of FAS at Cambridge, and I offered to help. Later on I established a chapter subcommittee on radiation hazards, which had the related effects of getting my husband to propose to me and helping to stop testing of atomic weapons in the atmosphere.

In the fall of 1954, Aage Bohr came to give the Loeb Lectures at Harvard University. Aage is the son of Niels Bohr, who devised the first model of the atom and who taught most of the giant figures in modern physics. He is a theoretical nuclear physicist whose work on the collective model of the nucleus would win him a Nobel Prize. His wife, Marietta, came with him. I had gotten to know them both through the Lauritsens, who were their close friends. Marietta and I had hit it off immediately and she had become a dear friend. She had also come to the States as a refugee, and had met Aage when he accompanied his father to Columbia University. Marietta was a very intelligent, sensitive, spunky, and determined person. (The word "determined" is surely an understatement.) One morning she came to have breakfast with me and said, "You have got to get married!" I agreed with her. She asked me what kind of a man did I think I wanted. I named a physicist we both knew. She told me that I was crazy and that I didn't understand myself. She said she had met the right man for me the previous evening at a party at the Bainbridges, an assistant professor at Harvard, Walter Selove.

I didn't know Walter but I had seen him at colloquia. He smoked during the lectures and always brought a small ashtray along. He seemed somewhat affected and arrogant. He definitely didn't fit my view of an attractive man. I had heard a great deal about the original and very creative work he had done as a graduate student at Chicago. He had designed and constructed a device called a neutron time-of-flight chopper to measure neutron energies and had published some beautiful results. He had been a graduate student of Edward Teller, which did nothing to endear him to me, and had worked at the weapons laboratory at Livermore for a year when Teller first set it up. I said "Nah" to Marietta, and she said, "You must find a way to meet him. Promise that you will do it, or I won't go back to Denmark." I did not think that this was a serious possibility, but I also knew that it was useless not to agree to this small request. Marietta went home. Shortly thereafter I saw that Walter was giving a colloquium at MIT on "Internal Momentum Distributions in Nuclei" and I decided that —why not?—I would go to listen to him. It would get me off the hook with Marietta.

And then something totally unexpected occurred. I sat in the front row of a large auditorium, and I fell in love with Wally while he spoke. I did not pay any attention to the scientific content of the talk. I just watched him and I knew that Marietta was right. I also found him very attractive physically and I wanted to go to bed with him. At the end of the lecture I went up to Walter and told him how much I had enjoyed his talk, and could I have a preprint of his work? And then for the next six months I conducted a campaign to get him to take me out. My tactics were sneaky and eventually successful. Wally was a member of the Federation of American Scientists. I asked him to chair a subcommittee to look into radiation hazards due to bomb tests. He agreed. I appointed myself to the subcommittee. I also made the mistake of appointing another woman. He started dating her. I bought two tickets to the opera, called him, and said that, gee, a friend of mine couldn't make it, would he, by any chance, like to use the ticket? "Sorry, no": he had another engagement. Several times I got myself invited by a couple who were friends of Walter: the man they invited to balance the genders was as little interested in me as I was in him. A well-known physicist came to visit from England. I had heard that Walter wanted to meet him. I asked my English friend whether he would mind being a shill, and explained. I called Walter and asked if he would like to come to my apartment to meet my friend. His response was "No": he had to do an experiment on the Harvard cyclotron.

Finally, somehow, I got through to him. When I returned from an out-of-town meeting on Memorial Day in 1955, he met me at the airport and took me to his apartment, and we made love. I was sure that it was love because I did not mind too much the fact that his apartment was dirty and filled with unopened boxes, nor even that the bed linen was of dubious color. We spent the next day driving around New Hampshire. Wally was in a good mood while I was sore, bleeding, cranky, and triumphant. I was twenty-nine and Wally was thirty-four years old.

Wally explained very carefully and repeatedly that he did not want to get married, and I was willing to accept that. Then my menses stopped. I went to New York to see Mother's gynecologist. He said, "Das ist eine alte geschichte" ("that's an old story") but concluded that I was not pregnant. I had thought things through before I heard the good news and had decided that I was not going to have an abortion, that I would not marry Walter, and that I would raise the child on my own. I did realize that to be an unwed mother would probably get me fired. There was no question in my mind that abortion was not a solution for me, personally. But I have always respected the freedom to choose that any woman must have in this most serious and private matter. I am appalled by the behavior of groups such as the Catholic church which oppose all abortions. No group has any right to make such demands of individual women, particularly a group such as the Catholic church whose behavior has been so often immoral in its fostering and acceptance of anti-Semitism, as well as in its view of women.

It was an upsetting period, and not helped at all by Mother telling Misha the whole story. Misha was furious and decided to shoot Wally, but I finally convinced him that it was I who had seduced him. In fact, it turned out that I was unable to conceive. We tried everything possible for the first ten years of our marriage. What I had experienced was probably a hysterical reaction to my first major "sin."

I already had commitments for the summer. Bill Havens, a professor at Columbia, had asked me to teach a summer session course in nuclear physics and to consult with his group on a possible future program in nuclear spectroscopy. Then the University of Mexico group wanted me to come and consult with them. Eugene Wigner had suggested my name, and they arranged to have the State Department appoint me as a Smith-Mundt Fellow. I had to report to Washington for a briefing. It was a formal and quite meaningless procedure. I remember that they told me to be careful not to drink the water and mentioned an excellent place to have milkshakes. With my usual tact I said that I preferred tequila. Then something indefinable took place. The man briefing me implied, or barely hinted, at the thought that, gee, perhaps it might be interesting for the U.S. to know something about the state of radiation fallout measurements in Mexico. I realized that, if I understood correctly what he was saying, it was inappropriate for a visiting scientist to dig up this information. I looked blank. Of course I was curious, so I found out for myself how hard it would be to obtain the data. All that one had to do was ask.

Sigrid Lauritsen, Tommy's mother, wanted to visit Mexico. So I dropped by Pasadena and Si and I flew on to Mexico City. I loved Mexico immediately. I liked the people, I was entranced by their culture, and I liked the tempo of the city. Si and I stayed at the Hotel Geneva, near the center of the city, and I tried to remember my Spanish. Heinz Barschall was also there because of a joint meeting of the American and the Mexican Physical Societies. We went roaming about the countryside and ate fabulous meals. The water did not affect me: in any case, Si was my private physician; but I was extremely lonely for Wally. Si was a wonderful and wise listener.

I had worked with one of the Mexican physicists at Buechner's lab at MIT. Bill Buechner had been very involved in helping Mexico start its nuclear program, and he had hosted Marcos Mazari for a year. Marcos was an engineer by training but he became a mainstay of the nuclear physics effort in Mexico. I had worked with him in Cambridge. Marcos is a marvelous man: a nice, intelligent, and very warm person with a large and delightful family. Unfortunately the program was being run on a shoestring budget, and it never really became competitive, though many excellent young scientists were trained at the Van de Graaff facility. My first shock came when I arrived at the facility and found that the vacuum pumps had to be turned on. The vacuum did not become adequate until the early part of the afternoon. At 3 or 4 P.M. most technicians and scientists disappeared. They had to hold second jobs to survive, and the vacuum pumps were turned off. This was because electrical power often went out during the evening hours, and the accidental shutting off of the pumps would have led to major difficulties. I suggested that they buy a small power plant of their own, and asked Misha to offer them one at cost, but that was probably too simple a solution. The second shock came when I went to use the library and found out that the books were catalogued not by author and title but by their date of accession. This did not make it easy to use the library. And finally, whenever I proposed giving one of the seminars I had so carefully crafted, I was told that tomorrow would be splendid. I don't remember giving a single talk, but the Mazaris took me to all the interesting sites near the city, and I learned to pronounce the names of the two volcanoes which overlook Mexico City. I also developed a great respect for Mexican-Indian cultures.

Both during the summers of 1954 and 1955 while I was away, I had invited friends to use my apartment at 100 Memorial Drive if they were in town. I left a list of names with the building manager. Anyone on the list could use my flat. In 1954 Sam Goudsmit did. I had become extremely fond of this interesting, cultured, feisty, and complicated man. As a young man he had invented, with George Uhlenbeck, the concept of electron spin. He came to the States as a refugee from the Netherlands. During the war he was the leader of a group called ALSOS whose mission was to try to find out how far Germany had gotten in building an atomic weapon. The answer was, fortunately, not far. This was determined by Sam and his hardy group, which entered Germany with our frontline troops. Sam knew many of the German scientific leaders from his years at the Bohr Institute before the war. He wrote a book which is fascinating even today.[8] Sam enjoyed my apartment. When I returned I found that It was appreciably cleaner than when I had left it, and that convenient kitchen utensils and interesting books had been added. We remained friends until his death. I have recently found several delightful letters from him. Sam had a large circle of interesting friends. One of them was Moe Berg, a Princeton graduate with an ear for languages and broad cultural interests who became a major-league baseball player and then a spy. He was also a physics buff. Sam introduced me to Moe, and we went out a few times. I enjoyed knowing Moe very much. We did not discuss baseball.

While I was still at Columbia during the 1955 summer, I called my apartment one day to find out who was using it. Denys Wilkinson answered. Denys and I had become friends while I was still at Wisconsin. He is an Englishman, a world-class electronics man, and a truly superb physicist: he devised the first pulse-height analyzer which is a dominant tool in all nuclear and particle work that has been done since the late forties. In fact, these devices also underpin much of technology. Had he been interested in money he would be a *very* wealthy man. Fortunately, he became a nuclear physicist and, in my opinion, is the very best of us. Denys also has very strong scholarly interests in bird navigation and baptismal fonts, subjects about which I became far more knowledgeable than I had ever expected to be when I prepared a talk for his retirement party in 1987. He has elegant and provocative ideas and is fantastically creative. He is also a gifted mimic and he has a notable sense of humor. His after-dinner speeches at physics meetings are legendary. He became one of my best friends. He was eventually knighted and, for a time, was Vice-Chancellor of the University of Sussex. His first wife, whom I did not know well, was a French physicist. His second wife, Helen, is a charming and very beautiful woman who is a singer.

Denys said that the apartment was delightful and that he used it only during the day when the cross-ventilation and the proximity to the MIT library combined to make it a wonderful place to work. However, he added, there *was* something odd. Every morning he found male shorts drying in the bathroom and a slightly lower level in the bottle of scotch. Did I know what was going on?? No, I didn't, but I called that evening and Wally answered. Wally had been sleeping there and he was curious to know why the apartment appeared different when he returned in the evening. I thought the whole thing hilarious and recounted the events to Bill Havens. Bill did not know that I had been living with Wally and he exploded when I mentioned his name. He said, "That bastard!" and wondered what a nice girl like me was doing with such a guy. It turned out that the neutron time-of-flight chopper that Wally built for his thesis made it possible to obtain results, at high resolution and with separated isotopes, which Havens's equipment was not able to do. This, in itself, is never the reason for a public outburst, although railing against adversity is okay in the privacy of one's home. Wally had referred to the

Havens results in terms that Bill felt were belittling, and
Bill was very, very angry. I have always wondered how
he reacted when he received our wedding announce-
ment a few months later, and whether he remembered
his outburst. I was amused and Bill and I remained
friends. But I don't think that Bill ever forgave Wally for
those unkind remarks. Bill became executive secretary of
the American Physical Society and was in charge of the
Programmes of the APS meetings. It became a joke in
our household that if Wally gave a paper, as often as not
it was scheduled for the end of the last day of the meetings, by which time most people had left and the attendance was miserable. It was probably not deliberate, but
it was certainly a bizarre coincidence.

I returned to 100 MEM in the late summer of 1955 and
resumed my life with Wally and my work at Boston
University. I was completely in love with Wally and
wanted to share my life with him. It seemed natural to
marry. Over the summer Wally had been lonely also and
he came to love me and to trust me. One Friday evening
he proposed; we spent the night making love and talking about the future. In the morning I called Viki
Weisskopf, to whom I was, informally, giving Russian
lessons and, announcing the great news, I told him that
today's lesson was not on. He was amazed that Wally
and I had become engaged because we had kept our
relationship quite confidential. Wally then said that he
wanted to buy me an engagement ring. I didn't care
about rings and I knew that he had very little money,
but he insisted. We went to one of Boston's best jewelry
stores, and a salesman in tails approached us. Wally said,
"We want an engagement ring." "Yes, Sir. In what price
range?" Wally, with great largesse, said $100. To his credit
the salesman said, "Yes, Sir," and showed us a gold ring

with a tiny diamond chip. Wally said this would not do, and we left the store. We finally found a ring in the right price range at a discount jeweler. It was really a wedding ring but it had half-a-dozen visible diamonds along its circumference. I couldn't have cared less. I was euphoric. On Monday I gave a memorably poor lecture to my BU students in which, in response to a question, I discussed semi-stable equilibrium.

I was no longer in equilibrium. I was full of joy. I sent a thankful and triumphant cable to Marietta Bohr, and I started to prepare for the wedding. Wally was insistent that the wedding be very small, that there be no announcements in newspapers, that I was to wear a long white gown, that we should cater it ourselves, that we were not to accept a large present from my family, and that a rabbi was to officiate. I almost balked at that last item but decided, as Henry IV had some years before, that "Paris vaut bien une Messe." We interviewed rabbis all over the Boston area to try to find one who would keep mention of the deity to a minimum. We would have been better off in a Unitarian ceremony. The rabbi we chose turned out to be full of previously hidden religious fervor.

Wally decided that once married we should live in a different apartment. He thought that there were too many other memories in our two flats. We found a dispiriting small apartment on Garden Street; to save money we immediately moved into it, and I had to give up my glorious view of the Charles. However, two of Wally's friends lived at 100 MEM and we decided to use the lounge on top of the building for our wedding. I spent the fall choosing a wedding dress. The dress had a train and was made of white velvet. Wally chose it over Mother's strong objections because a groom was not supposed to be involved in prenuptial arrangements. During the ceremony he wore sand-colored suede shoes.

I didn't care. I just wanted to get on with it and marry Wally. On December 18 (a day chosen to be at the beginning of the academic recess) we woke up, picked up my parents, Yvette and her husband, and Lou, one of Wally's two brothers, and drove to 100 MEM. We stood in front of our two dozen friends facing the Charles and the rabbi. Misha stood by my side and Lou by Wally's, and our two friends, Geoff and Rita Knight, were our best man and matron of honor. I was sorry that the Lauritsens had not been able to come, and that so many of our friends could not be invited because of the small space. But it was a pleasant ceremony, and we went off on our honeymoon after snacks and champagne. I look very grim on the picture that was taken as Wally and I left 100 MEM. This was only partly because Wally had offered to take his brother and Mel Gottlieb (later a professor at Princeton) along with us to New York, on our way to our honeymoon in the Florida Keys. Perhaps I was prescient and realized that Wally and Mel would whistle "Eine kleine Nachtmusik" for much of the four-hour trip.

When I first became serious about Wally, I tried to find out everything I could about him as a person and as a physicist. I sought out people who had known him at the Rad Lab at Cambridge during the war years, at Chicago, at Argonne where he had done the actual experimental work for his thesis, and at Livermore, as well as at Harvard. The picture which emerged was of a quite wonderful man and a superb experimental physicist. He had sought the appointment at Harvard, instead of accepting any of several other good offers, because he felt it better for his personal life. But he did not have the political sensitivity and savvy to advance at Harvard; I knew he would not receive tenure there. What I had consciously wanted to understand was whether Wally was a better physicist than I. If I could respect him, and I could, then he would respect himself, and then I could be free to be a physicist and to lead a demanding professional life. And I still wanted that, more than ever.

Wally's parents, Rose Feld and Abraham Yoselovich, had emigrated to the United States at the beginning of the century from eastern Europe. They met in the States and had five children, four boys and a girl. One of the boys died young. Wally was the youngest of the boys and older than his sister. His parents changed their names before he was born. When he was growing up in the thirties in Chicago, the family was devastated by the Depression. His father, a peddler and house painter, had great difficulty finding regular work. For a time the family owned a fruit and vegetable stand, operated by Wally's parents and his two older brothers. One of the older brothers, Joe, became the protegé of the wife of a University of Chicago professor, who financially aided him to go to the university, where he also received a Master's degree. The other brother, Lou, took a job in Washington and for many years helped to support the family. Lou was a fireman for a while and encountered a great deal of discrimination. He eventually served on the ship *Exodus*, which attempted to bring Jewish refugees to British-controlled Palestine; then he lived on a kibbutz in Israel, where he married an American woman and returned with her to the States. After some time he went back to Israel, where he died a few years later. His last job was as a draftsman and furniture designer. Joe married a gentile woman from Kansas, and they raised nine children, on a farm with forty cows. Three of the children were Helen's, Walter's sister, who was unable to take care of them. Joe finally became a civil servant. He retired and is now dead but his children are living the American Dream: several are physicians and all are doing well.

Wally was treasured by his mother. He roamed around Chicago and helped earn money by hauling sacks of produce, making punchboards in a factory, and peddling geegaws at athletic events. He got beaten up by kids from other minority groups and was called "kike." He began to build radios, practiced Morse code, and "corresponded" with other radio hams all over the world. He bought an old Model A Ford for twelve dollars with a friend; it came with an extra set of piston rings. He learned to take the engine apart and put it together again, increasingly quickly and easily. (This is the kind of invaluable experience which youngsters no longer have in our world of unopenable black boxes; and computers do not serve the same purpose.) He and his friends pooled their resources to search wrecking yards for spare parts and on good days to buy a little gasoline. Wally was a superb student in high school, and the University of Chicago offered him a tuition scholarship.

The university awakened his intellect. He had thought he wanted to be a radio engineer but Chicago has no such major. He was advised that physics was the closest area. Wally sought out extra work to do in his classes because he needed an A average to maintain his scholarship. He found physics to be much more interesting than he had imagined, and he was fascinated by marvelous courses in the classics and in history. He learned about the world of ideas, and he began to understand what was happening in the United States and in Europe. He lived at home and one year, to help, he worked full-time during the day in a factory, taking classes in the evening.

Wally has central vision in only one eye because of lack of medical care when he was small. He also has a severe case of Crohn's disease, coupled later with a botched operation in a public ward at Chicago which had long-term severe after-effects. (Crohn's disease is also known as ileitis. It produces chronic inflammation of the intestines and may result in intestinal blockages.)

When the war came, Wally first taught electronics to Signal Corps students at Chicago and then went to the Radar Lab at MIT. His work there, on moving-target indicators using radar, led to a number of patents. After the war he returned to Chicago as a graduate student, with a National Research Council Fellowship. Both his parents had died. Wally arranged to work on his thesis with Edward Teller. Teller suggested that he do a theoretical analysis of resonances observed when neutrons are scattered or absorbed by nuclei. After looking over the experimental results, Wally decided that they were so poor that an analysis would be fruitless. So he decided first to get better data, and developed the neutron time-of-flight chopper which provided a major advance in the field.

In 1955 the life of a woman physicist was very rough. When she heard that I was getting married, Maria Goeppert-Mayer, who saw me at a meeting, invited me to her room and poured me a stiff slug of whiskey. She told me that while it was hard to be a woman physicist, it was nearly impossible to be a married woman physicist. This distinguished woman, who was later a co-winner of the Nobel Prize with Eugene Wigner in 1963, had not been able to get any regular academic jobs in the United States. Most women scientists marry other scientists. "Nepotism" rules were used by universities to exclude women: if their husband was on the faculty, they could not be. Maria's husband, a very well-known physical chemist, was on the faculty at Johns Hopkins and then at Chicago. She taught courses, did research, and supervised doctoral theses there without pay. She became a professor at the University of California, San Diego, only two years before receiving the Nobel Prize. A second example was C. S. Wu, who *should* have been a co-winner of the Nobel Prize with T. D. Lee and C. N. Yang in 1957. Lee and Yang had done the theoretical work which she showed, in a magnificent experiment, to be correct. At least this accomplishment finally landed her a professorship at Columbia. (Her husband, Luke Yuan, was on the staff of Brookhaven National Laboratory.) In time, Chien Shiung Wu became the first woman to be elected president of the American Physical Society. But in 1955 life was very rough.

Wally was about to start on a one-semester sabbatical leave and had arranged to spend it at Brookhaven, to begin work in particle physics. He had also applied for a senior NSF postdoctoral fellowship, and he received it shortly after Harvard decided not to grant him tenure. With the fellowship, he resigned from Harvard. I was on the verge of obtaining tenure at Boston University, but we needed two permanent jobs. There were only two possibilities. First, the University of Minnesota offered Wally an associate professorship, with tenure, and me either a teaching or a research assistantship, the type of appointment usually given to a graduate student. But we chose our second option. The University of Pennsylvania offered an associate professorship to Wally, and there was a school nearby, Haverford College, which was interested in me. Aaron Lemonick was chairman of the three-person physics department at Haverford. I had met him during the times I used the cyclotron at

Princeton, where he was a graduate student, and we had become friends. He recommended me to Louis Green, who was essentially in charge of all sciences at Haverford, and Louis interviewed me. I stayed overnight with Louis and his wife Elizabeth in their apartment in an old house on that beautiful campus. They were the first Quakers I had met, and I enjoyed their calm while eating a simple dinner presented on beautiful china by a maid.

At the time, Haverford was strongly dominated by the Society of Friends and attendance at Quaker Meeting was expected of faculty and students. I explained carefully to Louis that I was a Jew and an atheist (to which he mused "most of us are, you know"), and that I would not attend the weekly Meeting (in fact I did, a few times, but I don't like to feel coerced); I also explained that I was definitely not a pacifist. Nevertheless, he hired me. I was to become the first woman to work full time on the faculty of Haverford, which was an all-male college at the time. Both the Penn and the Haverford offers had to be in our hands at the same time before we could accept them. There were some delays but we were told not to worry. The chairman of Penn's department was Wilbur Ufford, who happened to be Louis' brother-in-law. My parents were strongly opposed to my going to Haverford. It was a job but not one that was likely to permit me to remain a professional physicist. I knew I had no other options, but I hoped to find ways of remaining active as a physicist.

In the spring of 1956, while Wally was working at Brookhaven, I taught at Boston University, living on Garden Street and disliking every minute of it—100 Memorial Drive had been far more pleasant! In the summer we went to Europe for six weeks: Wally had finally agreed to accept this wedding present from Misha. We had a glorious time. Wally, who had never been to Europe, met Sara and they liked each other immediately. We traveled to places I knew, as well as to Italy and Denmark. We dropped by Monte Carlo where I kept tossing dice and winning at roulette. I found gambling extremely boring, but we made enough money to make this part of the trip free, plus I bought myself a Hermès scarf and we had a splendid meal at the casino. We sat on stone steps in Florence and listened to Tebaldi. At Wally's insistence, we had rooms with private baths only every other day, to save money. I was used to public facilities but Wally discovered that private bathrooms were decidedly more pleasant. In the fall I joined him at Brookhaven and supervised my students' work at Boston University by phone and through brief trips. We also commuted to the Philadelphia area on weekends to find a house that we could afford. We finally bought a very pleasant four-bedroom house on Brookview Lane, in Havertown: it was set on a quarter-acre plot, on a wooded hill, close to Haverford and near public transportation to Penn. It cost $22,500, which was a fortune for us, but Misha helped with the down payment.

We were ready to begin our real life together.

Fig. 1. My father, M. A. Ajzenberg, circa 1910. He is wearing
the uniform of a student at the
St. Petersburg Mining Academy.

Fig. 2. My mother, Olga
Ajzenberg, circa 1909.

Fig. 3. My father Misha and I, circa 1936, in Paris.

Fig. 4. Our wedding at 100 Memorial Drive, Cambridge, December 18, 1955. *Left to right:* Lou Selove, Wally, me, my mother, Misha, my sister Yvette, and Yvette's husband Ziutek.

Fig. 5. Teaching at Haverford College in 1960. My students are, *from left to right,* Harold Taylor, Curtis Callan, Samuel Tatnall, and Robert Cornwell. (Photograph courtesy of Haverford College.)

Fig. 6. Atomic Energy Commission delegation to the USSR,
February 1966, Moscow. *Left to right*, M. Abrahams (AEC), two of
our Soviet hosts, and J. L. Fowler (Oak Ridge) in back; then J. R.
Huizenga (U. of Rochester), George Rogosa (head of the delega-
tion, AEC), me, H. H. Barschall (U. of Wisconsin), Vance Sailor
(Brookhaven National Laboratory), James Leiss (National Bureau
of Standards), Louis Rosen (Los Alamos National Laboratory).
(The picture was taken by Richard Diamond of Lawrence
Berkeley Laboratory, to whom I am indebted for permission to
reprint this photograph.)

Fig. 7. Tom Lauritsen at Caltech, 1969. (Photo courtesy of the Archives of California Institute of Technology.)

Fig. 8. At the meeting on "Women in Physics" in New York, January 1971. *Left to right:* C. S. Wu (Columbia University), me, and Betsy Ancker-Johnson (now at General Motors). (I am indebted to Gloria Lubkin and to *Physics Today* for permission to publish this photo.)

Fig. 9. Herman Feshbach, W. A. Fowler, F. Reines, and H. H. Barschall, at the fiftieth birthday party of the Kellogg Laboratory at Caltech, 1981.

Fig. 10. Margaret (Lauritsen) Leighton and her husband Robert Leighton at Caltech, 1981.

Fig. 11. My husband, Walter Selove, in St. James Park, London, December 1985.

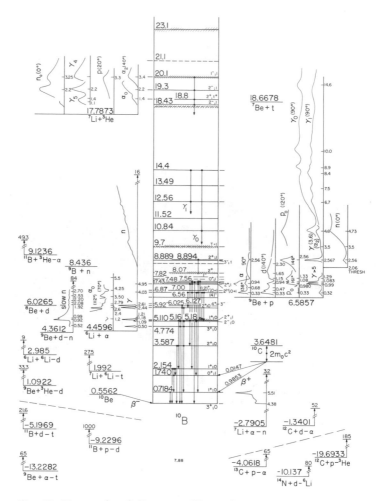

Fig. 12. Energy level diagram of B–10, from F. Ajzenberg-Selove, *Nuclear Physics* A490 (1988): 1–255. (I am indebted to P. D. Greaves, J. Visser, and the Elsevier Science Publishing Co. for permission to publish this picture.)

4

Haverford

1957–1970 ················

THE HOUSE ON BROOKVIEW LANE was a delight. A carpenter built bookcases for us and we started unpacking. I converted a small room next to the kitchen, overlooking a green, leafy ravine, into my office. Wally had a larger den upstairs. We had a proper guest room and two bathrooms. I became house-proud. And a wonderful thing happened. I had been looking for a cleaning lady to come in one day a week, and Edith Briggs took the job. Edie is a remarkable lady. For the past thirty-six years she has taken care of us with good humor and with enormous thoughtfulness and kindness. Now in her mid-seventies, she is still our part-time housekeeper. Edie is one of the most responsible, moral, and intelligent people I have known. Because she was born black in a racist country, she has not had the opportunities she deserves but she is a proud and strong woman. I respect and love Edie.

The physics department at Haverford was located then in a beautiful and inefficient old stone building. There was no ladies' room. I said that I didn't care, that I could share the men's toilet. After my days at the Jungfraujoch, any unalarmed toilet was fine with me. But Haverford promptly built a toilet for me. The laboratories took longer. Aaron Lemonick had been at the college for only a short time. We found to our dismay that there was virtually no useful demonstration or student laboratory equipment available; nor was there a technician to build anything that we might design. We began to write proposals to obtain equipment funds from the college and from the National Science Foundation. Then Aaron received an offer to return to Princeton and, to my great sorrow, he did. This was good for Princeton, where he served as professor and dean of the faculty, but terrible for the college and for me personally. Aaron was the best classroom teacher I have ever met. The students floated out of his classes and worshipped him. I still like him very much. He is an absolutely honest person with common sense and charisma, and it was a joy to work with him. He also protected me from the back-biting faculty, relatively few of whom were involved in scholarly activities and therefore had more time to engage in petty squabbles.

I was able to transfer my grant from the air force to the National Science Foundation, making my Quaker colleagues more comfortable, and I continued my research program in nuclear structure physics, as well as my work with Tom Lauritsen. Over the first seven years at Haverford, I wrote half a dozen research papers, based on work done at accelerators at Princeton, Oak Ridge, and Los Alamos, and three review papers with Tom. I involved the undergraduates majoring in physics in research, both during the academic year and the summer, and they published papers with me.

The students at Haverford ranged in ability from good to brilliant. At the time they were primarily the sons of professional families. They came to Haverford rather than to one of the top Ivy League schools because they, or their parents, had decided that small classes and a great deal of attention from their teachers were preferable to diversity and the opportunity to interact with many first-rate scholars. It was a joy to work with them, although some of the classes had too few students to permit interesting discussions. Three of my students have become members of the National Academy of Sciences, and many others are having distinguished careers in science and in medicine.

I recently found a letter from Joseph Taylor, university professor of physics at Princeton, written when he was a sophomore. He asked me whether I thought that he was good enough to become a physicist. Fortunately I replied "Absolutely." Joe recently received the $100,000 Wolf Prize for physics for his work on binary pulsars. He has a fair chance at a Nobel Prize in a few years, and I shall return the letter to him then, with enormous pleasure. Another student was Curtis Callan, a world-class theorist, who is also a professor at Princeton. Curt, who came to Haverford at fifteen and graduated summa cum laude at eighteen, went with us on summer jaunts to Brookhaven National Laboratory on Long Island and to the Aspen Center for Physics in Colorado. Two of my present colleagues at Penn, Brig Williams and Richard Van Berg, who have been prominently involved in the search for the top quark at Fermilab, were also students at Haverford, as was William Forman of the Harvard-Smithsonian Center for Astrophysics who, together with his wife Christine Jones, was awarded the Rossi Medal in 1985.

One of the things I liked best about Haverford was its honor system. We announced the conditions under which a particular paper or test should be taken, and we let the students get on with it. Exams were never proctored. For instance, we could give a take-home exam and tell them not to take more than three continuous hours to complete it and not to consult any books. And we were certain that there would be no deviation from the rules. In my thirteen years at Haverford, I knew of only two cases where the honor system was abused. In the first a student had used a book. A friend of his asked him to report himself. When he did not, his friend reported this to the Student Committee on the Honor System, which took appropriate action. In the second case, I had suspected cheating but had no proof. The fellow in question was under severe psychological stress, and called me several years after graduation to tell me of his remorse.

Another excellent Haverford tradition was to have seniors write theses. Two come to mind. One of my first students was Frank Dietrich, who went on to get his Ph.D. at Caltech and who is an excellent physicist and a group leader at Livermore. When I met him he was about nineteen years old and determined to become a nuclear engineer. I thought Frank was so bright that he should consider doing physics research, but he was stubbornly insisting on nuclear engineering. I called a friend, Clark Goodman, who was an official in the reactor division of the Atomic Energy Commission (now the Department of Energy). He loaned me 5,000 pounds of uranium, consisting of 1264 slugs, in aluminum canisters.

Frank wanted to design a subcritical reactor. The three basic elements of such a device are the fuel (the uranium), a source of neutrons, and the moderator, a substance used to decrease the energy of the neutrons so that the fission of the uranium takes place at a reasonable rate. The moderator was easily available—it was tap water. The neutron source, a mixture of the elements plutonium and beryllium, could be bought. Frank designed the most efficient geometric array of the uranium slugs, and how to suspend them, by stringing them on aluminum rods, in the water. The water required a container, of course. The cheapest container was the kind of wooden barrel in which olives are shipped.

We agreed that we could allot twenty dollars for the barrel and Frank went off to the docks of Philadelphia to buy one. Before he left, I suggested that he measure the width of the door to the lab, and Frank, as usual, sneered slightly at me. When he returned he had a barrel which was too large to get through the door. He became very quiet and took the barrel apart. It consisted of seventy-six staves, and when he reconstructed the barrel inside the lab, it leaked. He spent most of a summer inside that barrel, caulking it, painting it, and cursing. Then one evening, around 11 P.M., he called me and said, in a controlled voice, "The neutron counts are going exponential. What do I do now?" In unison, Wally and I burst out laughing and said, "Run!" It is not possible to design a reactor using ordinary uranium and tap water which will lead to a critical chain reaction, with the number of neutrons increasing exponentially with time. The counter with which Frank was testing his array had developed a malfunction. The work Frank had done was remarkable, particularly so given his youth and the lack of resources. It was published by him in the *American Journal of Physics*. Unfortunately the local fire department heard the word reactor and announced, in the local newspaper, that it would no longer fight fires at Haverford. That flap was solved, but we were left, after Frank graduated, with 5,000 pounds of uranium, for which we had to account twice a year to the AEC. Frank has remained one of my closest friends. He is a wonderful man, and we are, unfortunately for him, saddling him with the job of being one of the trustees of our estate.

Another excellent thesis was completed by C. L. Cocke. Lew, who is now professor of physics at Kansas State University and a recipient of the Max Planck Research Award, was interested in studying the collisions of protons with protons. Bubble chambers were then an excellent detector for such events. In a bubble chamber, electrically charged particles create a disturbance which, when photographed, appears as a line on the photograph. Wally got permission for him to work at the Cosmotron accelerator at Brookhaven. Lew rebuilt a small commercial propane bubble chamber and designed the equipment that would trigger it and automatically take photographic pictures of the interactions when they occurred. Unfortunately there was no way to test the design before the actual run. The run had to be completed in four hours during one night, and while Wally watched, Lew did it. He patiently hand-triggered the equipment every eight seconds, and his results, and their analysis, were very pretty. My students published a number of other excellent theses in various physics journals.

In the meantime, W. E. Stephens, a professor at Penn, had decided that the University of Pennsylvania should have a program in nuclear structure physics. I had known him for several years, and he invited Haverford to join the university in requesting funding for an advanced type of Van de Graaff from the National Science Foundation. My undergraduates and I used the accelerator, and we wrote a number of papers in collaboration with staff members at Penn. Bill Stephens also asked me to suggest the names of good senior physicists who could work in this program. I suggested two names. One of these was an English physicist, Roy Middleton, who became the senior person at the facility.

We had moved to an elegant new building at Haverford. Several of us had been involved in its planning. Since the Board of Managers of the college was extremely interested in the project, there was now an abundance of well-designed space for instructional laboratories; and all faculty members in the three science departments which shared it had space for research activities. The library was excellent, and a pleasant auditorium was used for teaching and for guest lectures. We even had a part-time technician and a part-time secretary.

Perhaps the most important component of our program was that we had funds to bring in outside lecturers. An alumnus named Phillips had left a large sum of money and a collection of erotica to Haverford (the latter was promptly sold by the college). Half the income was to be used to invite distinguished "scientists and politicians." Typically we had two very eminent physicists every year as Phillips Lecturers. We were able to offer them excellent honoraria and they spent several days lecturing and talking informally with the students. They were a breath of fresh air in an atmosphere which would otherwise have become stiflingly self-centered. Many of the most prominent American physicists came to Haverford. They included Dick Feynman, Willy Fowler, T. D. Lee, Viki Weisskopf, and many others. Their influence helped the students to be more worldly—I was always concerned that they become aware of how physics is really done. At first I cajoled my friends at government labs and at universities to take them on as summer students. After a while cajoling became unnecessary because the kids were doing so well. Wally and I also took students along with us on trips to meetings of the American Physical Society.

There were many interesting visitors at Penn as well. Wally invited Edward Teller to come. Edward was persona non grata at the time with most physicists, following the Oppenheimer affair, and many Penn people were not friendly. He stayed with us—originally over my dead body—after Wally pointed out that it was his house too. As soon as I met him, I became fascinated by Edward, and I feel enormous respect and much affection for him. I think that he has contributed greatly to the security of the United States, and he has a brilliant mind, spewing out ideas at an enormous rate. He also has great charm and charisma. I was so enchanted by him that I washed and ironed one of his shirts. I do not usually do laundry.

From time to time I kept track of the way in which I divided the over ninety hours per week that I worked. Roughly, I spent one-third each on teaching, preparing lectures, and talking with students; carrying out my research and scholarly work; and doing committee work at the college and in my field. The committee work at Haverford could take up all of one's time, and some professors, it seemed to me, did just that. We were involved in virtually all decisions, with the president—then a Japanese scholar, and a very fine man, named Hugh Borton—anxious to hear all views. Meetings were conducted by the Quaker sense-of-the-Meeting procedure. No binding votes were taken, but as everyone discussed the matter at hand, and gave their views on it, it became evident whether a consensus was being achieved. We generally listened closely to each other, and if a minority could not be convinced that a majority view was correct, then, sometimes, the majority view changed. At first I thought this kind of procedure was madness, and certainly some of the committee meetings were achingly long, but I came to respect the process. It meant that no one was a clear loser or a clear winner. In the long run, this way of doing things decreased friction. On the other hand, the system was poor if cliques existed, and they became increasingly important as the Vietnam War polarized the campus.

My first few years at Haverford were extremely pleasant: the students were superb, the faculty reasonably congenial, and the campus was extremely beautiful. There was a sense of civility and of intellectual discourse. But the war in Vietnam changed that, as it changed so many other areas of American life. Most of the vocal faculty members were politically "liberal" and very much opposed to the war in Vietnam, as indeed were many of the students. There were no explosions of the type that occurred at Columbia and other universities because there was so little disagreement at Haverford. I considered Vietnam to be a sad but relatively minor war, though surely not to the people who were a real part of it. It became, of course, traumatic for a generation of Americans, those who fought in the war, as well as those who did whatever they could to resist fighting in it, and to their families. It brought to the fore class distinctions in the United States: a college student was far less likely to be drafted.

In some ways, Vietnam had the effect on America that the Algerian war had on the French: it showed the limits of the power that an industrialized nation can have on a Third World country. The long-range consequences of these wars have probably been beneficial for both France and the United States, both in establishing geopolitical limits and in bringing talented refugees to the respective countries. It is unlikely that the large number of Vietnamese who came here, nor the Pieds-Noirs—the Franco-Italian refugees from Algeria—who immigrated to France, would have been permitted to settle in the two countries without the bad conscience which we and the French had developed as a result of the two wars.

When Aaron Lemonick left, the college hired a very bright mathematical physicist to replace him. Unfortunately, this colleague mounted every barricade in sight. He even left for Vietnam to demonstrate against the war,

without asking me to take over his classes—or even cancel them—and without telling me to make departmental
decisions. When I berated him upon his return, he explained that he had not wanted the government to stop
him. It seemed to me that he and his many acolytes at
Haverford had a well-developed sense of paranoia, and
no common sense. The political cliques formed during
the Vietnam years greatly influenced the governance of
the college for a number of years.

In the meantime, Misha had died. On May 13, 1962,
Mother called me. She was visiting my father at his cottage in New Jersey when she heard him fall in the living
room. He was instantly dead of a massive heart attack.
Wally and I drove to New Jersey and, together with
Yvette and Ziutek, we did what was necessary. After the
funeral, Mother came home with us but she quickly decided that she wanted to go back to her New York apartment. Misha had chosen Yvette and me to be executors
of his estate. He had left half of it to Mother, with the
residue to go to Yvette and me upon her death; and he
divided the remainder between Yvette, me, and his relatives in Israel. Unfortunately matters quickly became very
complicated. Misha had sold his company to an Iowa-
based firm with the written agreement that if that firm
did not fulfill certain conditions, the company would
revert to my father. This reversion did, in fact, take place,
and Misha then sold his company to a conglomerate. His
estate principally derived from the proceeds of this sale.
When he died, the Iowa people apparently decided that
they could trample on three women. Their lawyer filed a
suit in Iowa accusing Misha of fraud, and demanding
essentially the entire estate. I was incensed because Misha
was a totally honest man, and I was proud of him. I
would have been happy to shoot the lawyer.

But first we had to fight the case and win. Mother thought that fighting a case in Iowa when the beneficiaries would be Iowans would be impossible. She had no trust in the fairness of the judicial system. I trusted the justice system somewhat and, in any case, I was not going to let the calumnies about my father remain on the record. I decided to fight, and Yvette finally agreed, though Mother said paying our lawyers would be costly and would lead nowhere. She refused to speak to me for six months or so. But we went and fought, and two years and an incredible number of wasted hours later, the court decided in our favor. I then wanted to sue the lawyer for libel, and found that in some ways the law is an ass: it is not against the law to libel a dead man. I hope that this lawyer's children came to be aware of his dishonorable behavior, and that they are ashamed of him.

This occurred at about the same time that I developed a seven-year itch about Haverford. In 1964 I was awarded a Guggenheim Fellowship, which I used to do research at the Lawrence Berkeley Laboratory. It led to a reawakening of my joy in doing intense research in nuclear physics. The award of the fellowship was also a great boost to my morale. I had been promoted to professor of physics at Haverford in 1962, but only after some trauma. Louis Green, who was extremely influential at the college, seemed to be pleased with the work I was doing, but he kept saying, "Gee, it would be nice if there were some research going on in the physics department." What he probably meant is that it would be great if there were research hardware *in situ* with which the kids could work, and I agreed, but clearly I, a nuclear physicist, needed very expensive equipment to do original research, and so we had to be users of other people's facilities. Still, we *were* doing research, and both my students and I were publishing articles and giving talks at national and international meetings, and our work was recognized and we kept on being funded by the government. Louis's repeated comments may have contributed to the climate that led me to quit Haverford in 1969. I do not think that Louis was conscious of what he was doing, although when I left he asked me whether any of his actions had been involved in my decision. I felt sorry for him, and said "No." But I had begun to realize that there was a problem when I received the Guggenheim in 1964. I had applied for two fellowships, the Guggenheim and, since I did not think that I would get it, I also applied to the National Science Foundation for a senior postdoctoral fellowship for college teachers. For the latter, which was less prestigious and which involved my work as a teacher, I asked Louis to be one of the referees. For the Guggenheim, I listed referees who could speak to my research potential. I did not receive the NSF Fellowship; and it appeared to me that Louis was startled when I told him that I had been awarded the Guggenheim Fellowship.

At the same time, our lives were changing in many other ways. Wally was setting up a particle physics group at Penn, and he was actively involved in changing U.S. policy on open-air testing of nuclear weapons; and I was becoming very interested in science policy questions.

When Wally began his work at Penn, there was no group in particle physics, although three members of the department were in the process of switching from nuclear physics to that field. Wally began to build up the necessary infrastructure, to hire very able technical people, and to train them. Many of these people are still at Penn. The expanding particle physics group also hired four younger faculty members as assistant professors. As a physicist, this all sounds great, but as a wife, very much in love with her husband, this was, on the whole, a pretty dismal period. Wally worked hundred-hour weeks, and was very often out of town, using the big accelerators at Princeton, Brookhaven, Stanford, and Carnegie-Mellon. During one lonely Christmas–New Year period he suggested that I accompany him on one of his runs at the C-M accelerator in a deserted area near Butler, Pennsylvania. We slept in a quonset hut near the accelerator together with a large number of cockroaches, and ate at fast-food joints. On New Year's Eve, the cyclotron beam was turned off for a few minutes and we danced and shared a bottle of champagne with the crew.

My mood improved when Wally and his team discovered a new particle. It was a heavy, neutral meson, and Wally was entitled to name it. He called it the f-zero, and it became known to a small circle of friends, and to all of my subsequent students, as the faon, after my name. Fortunately, the small circle of friends included Viki Weisskopf. When subsequent generations tried to bring some coherence into the naming of particles, and considered renaming the f-meson, Viki stuck up for the faon. It is a pleasure for me still to look up the *Particle Properties Data Booklet* used by physicists throughout the world, and find my particle there. It has now become the mother of a whole family of mesons.

I had appointed Wally chairman of the committee on radiation hazards of the Federation of American Scientists chapter in Cambridge in order to have a chance to meet him. When I first raised with him my proposal for a study of radiation hazards from nuclear tests, he made a quick calculation and said there was no appreciable problem. But I persisted, and he agreed to chair the committee. Wally is a serious man and he began a thorough study of radiation hazards in general, and, in particular, of the hazards due to atmospheric testing of nuclear weapons, which was then widespread. He studied everything that was available on the subject. Wally thought that the Atomic Energy Commission might have inadvertently erred in publishing data that indicated that the hazard to people was very low. He wrote to Willard Libby, one of the AEC commissioners, and Libby invited him to Washington to discuss these matters with him. Libby was very cordial, and Wally met with him several times over the next couple of years. He remained unconvinced by the government evaluations and wrote reports and letters to the editor of the *New York Times* and the *Washington Post*. I had no question that Wally was correct but I thought him mad to fight the government. I had been taught to be cynical and I pointed out that Wally, then on the threshold of tenure and about to request research grants from the Atomic Energy Commission, was in a mighty vulnerable position. Wally told me I didn't understand the United States and went on.

Wally appeared at countless forums and testified before Congress. In fact, he helped to organize the May 1959 hearings of the Joint Committee on Atomic Energy of Congress on "Fallout from Nuclear Weapons Tests," and he edited an issue of the *Bulletin of Atomic Scientists* on that subject. A number of people were involved in this controversy, but I think that he had a special ability to inspire confidence. He was earnest, he thoroughly understood the data, he was not a liberal "crazy," he had worked at the Radar Lab during the war and, briefly, on atomic weapons at Livermore. I was very proud of his fearlessness and of his ethical compulsion to do the right thing. I should mention that his activities did not interfere with his receiving research grants from the Atomic

Energy Commission, which says a great deal about Willard Libby. He was a good man, who later married a friend of Wally's, Leona Marshall, a physicist who had worked with Enrico Fermi.[9]

In May 1957 Wally received a letter postmarked Penrhyndeudraeth, N. Wales, from Bertrand Russell. Professor Russell invited him to participate in a small conference in Canada, at Pugwash, Nova Scotia, "in order to make an appraisal of the dangers which have followed from the development of weapons of mass destruction." The meeting was underwritten by Cyrus Eaton, the chairman of the Board of the Chesapeake and Ohio Railway Company. Eaton extended a warm invitation to us both and sent detailed information on how to get to Pugwash, where he had been born. He also included a check for expenses.

Pugwash was an extremely pretty small coastal village, and the meetings, in which I did not participate, were interspersed with many social occasions. For the first time, scientists from both Communist and non-Communist countries discussed questions relating to nuclear survival, outside a strictly official arena. The participants on the Western side were *hommes de bonne volonté* without close government relationships. The Soviet and Chinese scientists were, of course, chosen by their governments. The participants were M. L. Oliphant of the National University (Australia); H. Thirring of the University of Vienna (Austria); G. Brock Chisholm, director-general of the United Nations' World Health Organization, and John S. Foster of McGill University (Canada); Pei Yuan Chou, vice-rector of Beijing University (China); A.M.B. Lacassagne of the Radium Institute (France); C. F. Powell of Bristol University and J. Rotblat of the University of London (Great Britain); I. Ogawa of

Rikkyo University, S. Tomonaga and H. Yukawa of Tokyo University (Japan); M. Danysz of Warsaw University (Poland); D. F. Cavers (Harvard, Law), P. Doty
(Harvard, Chemistry), H. J. Muller (Indiana, Biology),
Eugene Rabinowitch (the editor of the *Bulletin of Atomic
Scientists*), W. Selove (Penn, Physics), Leo Szilard (Chicago, Physics), V. F. Weisskopf (M.I.T., Physics) (USA);
and Academicians A. M. Kuzin, D. F. Skobeltzyn, and A.
V. Topchiev of the USSR. The predominance of physicists among the participants in this meeting was typical
of the epoch: they best understood, and were most
troubled about, the consequences of a situation in which
an increasing number of countries would come to pos-
sess nuclear weapons.

The USSR delegation was escorted by a man who was
their official translator. Even to our naive eyes he was
clearly a KGB type who, when Topchiev said something
that didn't precisely follow the party line, translated the
comments so that they did. Topchiev looked like a typical party apparatchik and, as vice-president of the Academy of Sciences, probably was, but I talked to him in
Russian and he relaxed, and I quite liked him. I beat him
at croquet and outdrank him at vodka. At the end of the
conference he presented me with his card and said, "Any-
time you want to come to the USSR just show this at the
frontier." Despite the seriousness of the subject, the first
Pugwash meeting was intellectually challenging and socially pleasant. The conferees stayed in three sleeping
cars of the Chesapeake and Ohio Railroad, which Cyrus
Eaton had arranged to have brought in, and in cottages
at Pugwash.

The Pugwash conferences still take place in other venues. They have become a grouping of well-meaning
people concerned about issues such as ecology, but that

first meeting was very special. Wally, and I, as his wife, also attended the second meeting, which began at Kitzbühel in Austria, and continued in Vienna, in the summer of 1958. The meeting in Vienna was quite extraordinary. The city put us up in hotels, we attended banquets in palaces and a ball at the top of one of the hills overlooking Vienna, and we enjoyed a quite fantastic evening at the opera. That was the last time that I saw the Soviet translator: he was most annoyed that he had been given a balcony ticket, as all but the most eminent of us had been, and he demanded, and finally received, a seat in the orchestra.

I began to travel to my own international meetings in 1957.

If you work on interesting science, it is essential to schmooze with other scientists in your research field, and find out what they are doing at the time they are doing it. Your own work will not be current if you wait until your colleagues' results are published, since this occurs at least several months after the work is completed. Attending meetings is essential. The meetings take place at venues throughout the world. They are organized in such a way that no one group has a recurring burden, nor a disproportionate influence, in running them. It is part of the fun of physics, and of science generally, that its practitioners become world travelers, well acquainted with international airports and with watering holes.

I think that if sociologists were to study the dynamics of the interactions between participants at scientific meetings the results would be very interesting. As a beginner I felt completely awed when I saw and heard the great scientists in my field; and I was duly impressed by the confidence with which my colleagues discussed their results. It seemed impossible to contribute to a field in which there were so many smart people, with so many wonderful ideas. Fortunately, scientists are, in general, helpful to young people.

Later, as a professional in the field, meetings become essential to display one's ideas, and to convince other scientists of their validity. Now, as an older scientist, I am amused by the startled looks of the young as they glimpse the name on my ID tag; and I listen, skeptically, to the latest scientifically fashionable ideas. But what seems to me to be sociologically interesting are the intense relationships between the people at scientific meetings. They are built on personal connections, on past scientific alliances and on rivalries. The latter are the result of struggles for recognition and for influence. The relationships between scientists are extremely intricate, and they are very important in determining whether they can do their research successfully.

My first meeting was at the Weizmann Institute at Rehovot in Israel in 1957. It was a very interesting meeting at which proponents of the shell model and of the collective model of the nucleus discussed the direction that nuclear physics research was likely to take in the future. Many prominent physicists attended the conference, including Wolfgang Pauli, a brilliant but unpleasant man. I still had my Air Force grant and the Air Force was willing to assist me, and some other American colleagues, in getting to Israel. It was a long trip, in those pre-jet airplane days. Wally drove me to McGuire Air Force Base in New Jersey. I flew on a MATS (Military Air Transport Service) flight to Paris, via the Azores, where we stopped at an air force base to refuel. My status was undefined: I was neither a member of the armed services nor a civilian dependent. Seventeen hours later, in Paris, I asked where the customs/immigration booth was located and was told to forget it, I was extraterritorial and I should just go to the bus. Then I flew MATS to the American base at Frankfurt and spent the night in Bachelor Officers Quarters. The next day the physicists being convoyed to Israel were put into two unpressurized DC3's, and since we couldn't climb over the Alps, we headed toward France, followed the Rhone Valley to Lyon, and then flew on to Rome. One of the DC3's no longer had functioning radio equipment, but the crew took it calmly. In Rome we refueled and went on to Athens, where we spent the night. The next morning we took off, and then we promptly landed again. One of the doors had sprung open, but since the plane was not pressurized this did not matter too much—except to the man who lost the jacket he had hung on the door. The remainder of our flight was uneventful. DC3's have remained one of my favorite airplanes, together with the "Connie," the Constellation with its pregnant basement lounge below the passenger deck. But I decided that I preferred to return to Europe in a more conventional way, on a commercial flight. I then flew to McGuire AFB via MATS and had a marvelous stop at the immense air force base in Keflavik, Iceland.

I had several Israeli friends: the Goldrings (Hanna G. had been a student at Boston University), the Talmis, and Amos De-Shalit. They were very hospitable and showed me the country. Igal Talmi took me to the places where he had fought and pointed out the Jerusalem hills that were not yet a part of Israel. We were welcomed by the president of Israel, and everywhere we were shadowed by soldiers with machine guns because of the possibility of terrorists. I did not wish to see Misha's many relatives, partly because I detest family "socializing" and partly because Mother loathed them. Unfortunately, because Israel is such a small country they seemed to know my daily activities better than I. They tried to reach me by phone, and I was called out of meetings on a couple of occasions. I like my Israeli friends, but the feeling that as a Jew I was expected to be pro-Israeli turned me off. I have never been a Zionist, and I have never been back to Israel. I wish them well principally because I feel a deep sadness whenever a Jew is killed by violence. Yes, it hurts me more when the person killed is a Jew, though I believe that I am otherwise unprejudiced.

In the early 1960s I became involved with the American Institute of Physics as executive secretary of a committee to study physics departments in colleges, and as a member of a committee to advise the AIP on how to conduct personnel studies. Through a friend from my Cambridge days, David Z. Robinson, who had become an assistant to the science adviser to the president, I served on a panel of the Office of Science and Technology. It was at the one meeting I attended at the Executive Office Building of the White House that I wore stockings. I was awed to be there, and I was also thoroughly uncomfortable because the chairs in which we sat around the meeting table were too high, and my feet dangled, not touching the floor. That work led to my becoming a member of a panel which discussed the "Impact of Research Funds on Education" at a hearing of the Research and Technical Programs Subcommittee of the House Government Operations Committee. I mention these things mainly because my activities bemuse me now as they intrigued me then. For me it was an opening to a sphere of life which was different from my regular life, and therefore very interesting. Now I know what I want and what I enjoy, and my life is far more circumscribed. And, also in the 1960s, I got involved, primarily with D. A. Bromley, of Yale University, in setting up the Division of Nuclear Physics within the American Physical Society.

As the number of physicists increased after the war, as the explosion of new discoveries took place, and, particularly, as federal funding of research increased, the monolithic American Physical Society became fragmented, driven by the needs of the subfields. I had heard from Wally that he and several other particle physicists had been discussing the desirability of forming a Division of Particle Physics. So several of us decided that it would be wise to go along with the tide, and we petitioned that a Division of Nuclear Physics be formed, to represent our field within the APS and, to an appropriate extent, to the science policy groups outside. I was on its first nominating committee. Leonard Schiff was our first chairman, and Heinz Barschall our second. Allan Bromley was the first representative of the Division on the Council of the American Physical Society.

Allan, who was President Bush's science adviser, has been a friend since he was a graduate student at the University of Rochester. We met regularly at physics meetings, and I enjoyed him and his wife Pat tremendously. Allan is one of the brightest people I know, with an encyclopedic knowledge of all of science, and a complete understanding of the way people behave. He is utterly candid and suitably profane. And Allan is also extraordinarily well organized and very thoughtful. He is a great guy. On my sixtieth birthday he brought me an original edition of a book by Madame Curie: we had run together on the Brookhaven Van de Graaff on my birthday some twenty years earlier, and he had remembered. Over the years we have disagreed on many issues, but he *is* refreshing. My favorite remembrance of him is of the time I was trying to do an experiment at Yale. He was director of the Nuclear Structure Center and he greeted me very warmly, which I took to mean that he hadn't read my latest diatribe against one of his science policy proposals. I got myself physically inside the tanklike equipment I was to use and was loading photographic plates in the dark when the building suddenly shook. I was not surprised when one of Allan's minions appeared and told me that Allan wanted to see me *right away*. I realized that he had finally read my manuscript.

As I mentioned earlier, I was beginning to feel alienated by the atmosphere at Haverford College. I greatly enjoyed my students, but I was dismayed by the college's politicized atmosphere in the wake of the Vietnam war; and I very much wanted to do research more intensively. The Guggenheim Fellowship made it possible for me to take a leave at the Lawrence Berkeley Laboratory. Our stay in Berkeley in 1965 was marvelous. From one of our friends we rented a house that overlooked the bay. I began working at the 88-inch cyclotron which was a premier facility in nuclear structure physics. The primary users were nuclear chemists, attached to the chemistry department at Berkeley, and staff members of LBL, many of whom were nuclear physicists by training. Actually there was no difference in the kind of science that both groups were doing but, unfortunately, there was a historical estrangement between the physics department at Berkeley and LBL. No nuclear physicists were members of the physics faculty. We seldom went to colloquia on campus, but many superb physicists, from all over the world, lectured on our hill. There was also an excellent library and an increasingly first-rate computer center.

One of the most fascinating people at LBL was Luis Alvarez, who received the Nobel Prize in 1968 for his

work in particle physics.[10] Luis was a premier experi-
mentalist. I knew him slightly through Wally, and we
once talked on a long flight from Chicago to San Francisco, and we chortled with joy over his page proofs of
the new book by Edward Purcell on electromagnetism.
It is such a beautiful treatment of the subject! I was entranced by Luis's mind and I listened to as many of his
seminars as I could. One of the best talks I have ever
heard was given by Luis at Stanford: it was a beautiful
description of his scientific sleuthing, showing that work
which had seemed to indicate that a magnetic monopole
had been found was incorrect. He deduced that the cause
of the erroneous results was a small detail of the experimental setup. I admired not only the way in which he
had found the flaw but also the extremely courteous way
in which he referred to the scientists who had been involved in the incorrect experiment.

I worked with Bernie Harvey, and later with David
Hendrie and with many other excellent people at the 88-
inch cyclotron at LBL. The facilities were splendid. The
technical staff took great pride in providing the experimenters with steady and "clean" beams. The detecting
equipment was state-of-the-art, and the views just outside the counting rooms were breathtaking. I made lifelong friends there.

In 1960 I had edited a couple of volumes on current
topics in my field of nuclear spectroscopy, and a number
of my friends and colleagues had contributed sections to
it.[11] The book made a quite important contribution to the
field. In 1966 a revised version was printed, but it was
clear that the field was moving so swiftly that a new
book had to be written: I persuaded Joe Cerny, one of
my LBL buddies and professor of chemistry at Berkeley,
to edit it. And Wally and I began a close friendship with

David Hendrie and his wife, Dixie. David was then a post-doc at LBL but he was older, mature, and savvy. He had been in the Navy and then got a Ph.D. at the University of Washington. He is extremely smart but people who do not know him are misled by his burly appearance and his wide-eyed questions. He is an idealist and a strong and very nice person, although he pretends to be gruff. When their two daughters were growing up I used to send them feminist records and books. He would try to needle me about it, but he failed. David is one of the few men I know whom I trust totally not to be able to discriminate on the basis of gender. Four others are Allan Bromley, Frank Dietrich, Herman Feshbach, and Walter Wales. The reasons why they are not sexist are that they are very strong men, comfortable with themselves, and that they are perfectionists. They will go for the best available person, regardless of gender. David and Dixie Hendrie have now moved east. He is director of the Division of Nuclear Physics of the Department of Energy.

When I married Wally I told him that I wanted to travel around the world. He said okay, if we had the time and the money. I began to save at once. In 1964 I developed a very severe case of viral pneumonia, and was taken to the hospital. On the evening of my admission, our doctor gave us 50-50 odds that I would make it through the night. I asked Wally whether he would take off six weeks to go around the world if I made it. He said yes. I fought through that night; and during the next few months while I was exhausted from the aftereffects of the illness, I made very detailed plans for the trip. We took it in the fall of 1965, while at Berkeley, and we flew first-class all the way. Our first leg was from California to Copenhagen, via a stop in Greenland. We had a wonderful time with Marietta and Aage Bohr and their children. Then we flew to Helsinki and on to Leningrad.

This was our first trip to the USSR. We were quite astounded by the incredible bureaucracy and inefficiency of the system, but delighted by the sights and by the people. Leningrad was a luminously beautiful city, but it had not recovered from the German siege during World War II. The buildings were crumbling and the people were drably dressed. In this pre-glasnost era it was also totally impossible to contact scientific colleagues: we were there as tourists, and our guides aimed to keep us that way. It was a major problem to be allowed to see the fabulous Impressionists at the Hermitage, which were then off the approved list. We took pictures of the apartment house where my parents had lived and of the Mining Academy. (Mother refused to look at either: she had come to loathe Russia.) Then we flew on to Moscow and saw the sights. Two are particularly memorable.

One was the sign in our hotel, in Cyrillic, proclaiming "There are no rooms, and there will not be any." Rooms were for hard-currency tourists like us. The second occurred when we were ready to leave from Sheremetyevo Airport, to take the night flight to Delhi. We were left there by our guide in a huge crowd of people, none of whom were being checked in for the flight. The flight was "delayed." I spoke Russian with a very pleasant official. We asked whether yesterday's flight had left. "No," she answered, and it was pretty clear that there would be no flights to Delhi for the next several days because of a large-scale battle between Indian and Pakistani forces. We became very worried. We found out our hotel had no rooms and our visas were expiring. We decided to get out of the USSR, in any direction, on the first available plane. But for that we needed to obtain information as to which planes had space the next morning.

The nice official gave us a voucher for the Aeroflot hotel nearby which housed airplane crews, and the U.S. Embassy gave us the home phone number of the Austrian Airlines manager. The manager was extremely understanding: it was hardly the first time that such an incident had occurred. We spent a few sleepless hours in the dingy Aeroflot digs. The next morning we flew to Vienna, with an unforeseen stop in Warsaw, which made me very nervous because I was suffering from a severe case of Middle-European paranoia: a naturalized citizen of the United States does not have full protection in the country whose citizenship she previously held, or in which she was born. There was no rational reason to be worried, but I still couldn't fully believe my good fortune in having overcome the uncertainties of my childhood. For many years after receiving a U.S. passport I carried it everywhere with me, even in the States.

Everything was possible in Vienna. In a few hours our tickets were changed. We flew to Athens, to Rhodes for a few days, and then to Beirut. There we switched to a PanAm flight which stopped briefly in Bombay and Delhi and, following the Himalayas to the north, deposited us in Bangkok, on the exact day for which we had hotel reservations. Then we flew on to Hong Kong, to Japan, and finally returned to Berkeley.

I continued to work at LBL until the beginning of 1966, and found out that I was about to be involved in another interesting trip. The United States and the USSR were cautiously exploring scientific exchanges, and the U.S. Atomic Energy Commission was preparing to send a delegation on low energy nuclear physics to the USSR, consisting of some nine scientists and an AEC official. Heinz Barschall was to be a member of the delegation and he suggested that I be asked to come also. I was exhilarated by this opportunity and I agreed to go. We didn't know until the last minute that the trip would actually take place, and the itinerary was going to be decided in Moscow, but I bought a cheap and very warm Chinese lamb coat and met my colleagues in Washington. First we had to be briefed. The only interesting thing I remember from this episode is that a State Department official told us how to flush a toilet on an Aeroflot plane and how to eat Chicken Kiev, which we would be having daily. The secret of Chicken Kiev is that it contains a great deal of melted butter underneath a hard crust, and if it is pierced, a stream of butter flies out like a jet. Thus we had to depressurize the chicken in such a way as to minimize the danger to our clothes, and those of our colleagues. The official was right: we did have Chicken Kiev every day, and it became harder and harder to remain serious when we poked the crust and smiled mischievously at those near us who were in danger of being buttered.

We flew to Paris and spent a night there. Mother was visiting Sara, and she escorted me to Le Bourget Airport. We took an Aeroflot flight to Moscow. I don't know whose bright idea it was to send us there in February. The temperature when we landed was -40. For us, as physicists, this was a first. At that temperature the Celsius and the Fahrenheit scales cross, and there is no need to state the units in which the temperature is given. It was *cold*, and during the next ten days, when it wasn't cold, it snowed. We were met by a man from the U.S. Embassy and by a couple of representatives of the USSR State Committee on Atomic Energy. We were invited to meet our ambassador, and we were told to keep in touch with our embassy while in the USSR. All mail from home was to be sent to the embassy, to then be transmitted to us. We were taken to a fancy hotel for VIPs, the Sovjetskaya, which was extremely pleasant by Soviet standards, and we began to make the rounds.

We visited laboratories in Moscow, Dubna, Kharkov, and Kiev, and then we split up, some of us going to Leningrad while others went to a previously classified lab at Obninsk. We were escorted by an official of the State Committee, a translator, and someone whom we called the shaman. Once, while we were waiting for a plane at Vnukovo Airport, he managed to get us into a lounge marked, in Cyrillic, "Reserved for Delegates to the Supreme Soviet." When I asked him how we qualified, he said, "Perhaps some day you will be one." Less pleasant was the time we couldn't fly from Kharkov to Kiev because of the weather, and he flashed an ID to railroad officials and bumped some traveling Russians from several overnight sleeping compartments. It was that night that I became forty years old. I celebrated it first on that train, sleeping—totally clothed—with three of my chums: Heinz, Joe Fowler of Oak Ridge, and Jim Leiss, who later became a high official within the Department of Energy. The next day we celebrated with a proper cake at our Kiev hotel. Wally's birthday greetings reached me several weeks later, at home. The diplomatic pouch had been delayed due to the weather.

On a typical laboratory visit, we were escorted to the director's office, then sat around a green, felt-topped conference table laden with juices, tea, and snacks. We listened to a long presentation of what was done at the laboratory, and then we were escorted to the lab. We were shown a number of reactors—that is, we were shown the shielding surrounding the reactor, and we dutifully climbed up and down the concrete blocks. I stopped wearing heels after a while. And everywhere we went, we were photographed. Toward the end of the trip I asked our KGB shaman to send me copies of any photos of me that might be good, but I never received any. I concluded that either I looked terrible in all of them, or, possibly, the photos became part of what must be a very thick file on me in Moscow.

We attended banquets, particularly at lunchtime, with too much vodka and enormous numbers of toasts to "peace and friendship." But the cold war was still on. At one of the labs I saw a Russian scientist whom I had met at a meeting in the United States. He warmly invited me to have dinner at his house with him and his wife. I responded that I would be delighted to come but that he should probably check with his people first. They said "no," which made us both sad. He brought me a few tulips to the Sovjetskaya from himself and his wife, having taken the Metro across town to get them to me. Still, I met many Soviet scientists, particularly at Dubna, who have remained friends.

On the plane home we wrote a fake abstract of a talk on our trip, supposedly to be presented at the next meeting of the American Physical Society. It was signed "The Kiev Collective" and a footnote gave our names and affiliations. This was too offbeat to appeal to our AEC colleagues and they vetoed it.

I returned home to a somewhat worried Wally, having decided never to drink vodka again, and resumed teaching at Haverford. Teaching was as wonderful as ever but I was changing and so was Haverford. I had an increasing interest in research and in working on professional committees. I was often asked to give lectures at universities and government laboratories, and I refereed many papers from physics journals and many government grant proposals. But I think that I would have happily stayed at Haverford, growing comfortably into a Mme. Chips, if it had not been for the retirement of Hugh Borton and the arrival of a new president, an economist with a public relations persona. Where Hugh was really interested in the views of the faculty, the new president seemed to me to listen, unfortunately, mostly to himself. In early 1969, disregarding my recommendation as acting chairman as well as the concurring recommendations of several other senior faculty members, he made a tenure decision which we thought would be damaging to the physics department. I asked him whether he really thought that we were wrong in not recommending tenure. I would have accepted his saying yes, because it is always possible that one is wrong. But he said no, he agreed with our judgment, but he couldn't stand the idea of having to tell the man that he was being fired. I asked him to reconsider, because his decision gave the wrong signal for continued scholarly excellence at the college. He told me I was pathetic. Of course I decided to quit.

I asked for a leave for the fall of 1969, at half pay, which I was due under the sabbatical leave program of the college and which it granted, and I went looking for a job. The best arrangement I could make was through Bill Stephens at Penn. I could be appointed research professor of physics, without tenure, and with my salary coming from my research grant. If I taught, the university would cover that portion of my salary. Bill said that he knew that I was a good teacher, and that there was no question that I could teach most semesters, on an ad hoc basis. I didn't care about tenure but I knew that I would be miserable if I couldn't teach. I was happy with the position. Roy Middleton told me that he looked forward to my coming to Penn and to my working with graduate students in his laboratory. I called the American Association of University Professors to inquire about the etiquette of notifying Haverford of my decision not to return after my leave. I was told that it would be proper to notify them immediately, since it involved a tenured position. I told the president that I would be quitting as of the end of 1969. He was astounded. Soon afterward, I was notified that the college would decrease my sabbatical pay to one-quarter of my normal salary. I replied that in that case I would cancel the financial gift I had pledged to the college, which more than made up the difference. It was unpleasant. I heard later that the president was shaken by my decision to leave, and that he behaved more circumspectly in his remaining years at Haverford. I am no longer angry at him. He inadvertently led me to the most productive and fun part of my life.

I was sorry to leave the wonderful students at Haverford and its beautiful campus; I still feed the ducks at the pond there. But I had to do it. During the fall of 1969 I worked to complete a number of projects and I made the transition to Penn.

5

Cherry Lane
1970–1984••••••••••••••

IN JANUARY 1970 I SPOKE WITH ROY MIDDLETON. My commitments, left over from my work at Haverford College, were at an end, and I had a number of ideas for Ph.D. theses problems. I was ready to start working with graduate students.

To my amazement Roy told me that he had decided that it would be inadvisable to have me work with graduate students, that there weren't a sufficient number of them, and that he would rather that they work with him and with the other regular faculty members. This was a shock but I could do nothing about it. Since I had always worked by myself, and with undergraduate students and colleagues at other facilities, I would continue to do so.

However, another problem quickly developed. I found out that teaching on an ad hoc basis was usually not possible because several of the regular professors had suffered decreases in research funding, and needed to teach. On one occasion I literally waited in the corridor until just before the beginning of class on the first day of the semester, waiting to see if the professor would show up. If he didn't, I could teach the course. Unfortunately for me, he showed up, which made me very angry because I had heard that he disliked teaching the course, and I would have loved to do it. And again there was nothing I could do. Bill Stephens was also dismayed and he tried in vain to persuade the astronomy department to let me teach elementary courses. I didn't blame them. I knew no astronomy, though I would have been happy to learn. I just needed to teach.

It was not a happy time. Tommy Lauritsen had developed cancer, and Wally was having severe difficulties of his own. He had been working at Cornell with an assistant professor from Chicago. The collaboration ran into serious problems, which were converted by the Chicago group into fatal ones.

Then in late spring I found a small lump in my right breast.

I was quite sure that it meant cancer, because all the women in Mother's family, with the sole exception of my sister who was not breast-fed, had breast cancer. I was taught to check myself once a month, and to take action immediately if I found something. So I went to see my gynecologist. At the time, mammograms were not accurate and mine did not detect the lump, which was just a couple of millimeters in diameter. My doctor suggested I wait. I had been told that two months was a likely doubling time for a cancerous tumor, and two months later the lump was about 4 millimeters. I went to the hospital of the University of Pennsylvania. I wanted to be operated on by Jonathan Rhoads, the senior surgeon. He was the father of one of my best students and the chairman of the Board of Managers at Haverford. He had been very kind to me when I was there. Dr. Rhoads was out of the country, and another surgeon, Julius Mackie, was recommended to me to determine whether surgery was necessary, and to perform it if it was.

Jack Mackie was great, and he later became a friend. He operated on me in September 1970. When I woke up in the recovery room I wanted to know whether the tumor had been malignant, but the nurse wouldn't tell me. I somehow thought to ask her what time it was. It was 5 P.M., and I knew that the procedure had started at 11 A.M., when I had been put to sleep. I understood then that it was cancer. My main reaction, at the time, was the joyous feeling that I had won over the bureaucracy. My next reactions were more sensible. The pain was very bad. Jack had performed an extended radical mastectomy, which meant that he had to remove tissue beneath the thoracic cage, a small part of the left breast, and some muscles of the right arm, which I had to learn, painfully, to use again. Jack is wonderful in all respects, but because he is concerned about drug addiction, he is very cautious about prescribing pain killers for his patients. Still, in the over two weeks I spent in the hospital, I learned how to begin to move my right arm again.

There had been no spread to the lymph nodes, and I was given four chances out of five of living for another five years. I was too weakened by the operation and by the trauma of the arm to think about the odds, but Wally was very hard hit by my illness and the possibility that I would die in my late forties. At the time, doctors did not realize that one had to worry about the people who love cancer patients as well as the patients themselves. Several people were extremely kind, but then there was the colleague who came to the hospital to visit me and announced, "Why all the fuss? Everybody has cancer"—a rather unhelpful comment to make to a forty-four-year-old woman. I did not remind him of that curious statement when he developed the disease later.

Heinz Barschall was wonderful to us even though a tragedy had occurred. Shortly before my operation, anti-Vietnam terrorists had bombed his laboratory—and those of several other professors at Wisconsin—destroyed it, and killed one of the postdocs. The terrorists claimed that the lab was doing war work. Besides the terrible loss of life, the destruction of the lab resulted in Heinz's essentially stopping his nuclear structure work at Wis-

consin. Despite the trauma, he came to the hospital to
see me, and he also sent us packets of research papers on
breast cancer so that we might decide for ourselves
whether I should get radiation therapy. We decided
against it. I went home, and Wally was incredibly tender. I could not bear to look at my enormous scars, which
took some time to heal. But they had to be cleaned very
carefully, every day. Wally stood in front of me in our
shower stall, and gently cleaned the wounds. On the
front of the stall he put up a sign, "Beautiful Faychen,"
in Russian. He also forced me to do the very painful and
very necessary stretching exercises so that I would learn
to fully use my right arm again. Mother, who had had a
similar mastectomy some fifteen years earlier, had been
living by herself and refused to do the exercises. She was
in pain for the remainder of her life. Wally drew a height
gauge on the wall, and each day I was expected to stretch
and reach a higher point. I have had very little residual
pain and my right arm is as strong as it ever was.

One of the other things that helped me to recover
quickly was our new home.

In 1968 we moved to a house closer to Haverford,
located in Wynnewood, a lush enclave within the Main
Line suburbs. The house on Cherry Lane has the equivalent of five bedrooms, a den, and, unbelievably, four full
bathrooms plus two toilets. It is set on a beautiful acre of
ground and overlooks a tree-filled ravine. I loved the
house immediately. It was relatively inexpensive: its previous occupants had painted every square inch of it a
dark pink. I sat on the floor of the living room for two or
three months while contractors came and went. At the
end the house looked bright and beautiful. We had even
redone the bath of the master bedroom to create a step-
down bathtub to remind me of Neubabelsberg (we raised

the floor, and sank the tub). We each acquired two dens and many more bookcases. And we had much more room to welcome visiting friends. It helped, during my recovery, that I was living in a place I so enjoyed. Six weeks after the operation, I was actually able to travel on my own to a meeting at Brookhaven.

Gloria Lubkin decided that I needed something to occupy me while I was still unable to work full time.

Gloria suggested to Brian Schwartz of the "Forum on Physics and Society" of the American Physical Society that I be asked to organize the first ever meeting on "Women in Physics," to take place at the national meeting of the APS in New York, in February 1971. I was interested in the notion. Besides my own relatively small problems, not all of which, surely, could be ascribed to my gender, I was appalled by the difficulties which brilliant women such as Maria Goeppert-Mayer and C. S. Wu had faced, and I had had a cathartic experience a few months earlier that had made me a feminist.

A young woman graduate student in physics knocked on my office door at Penn and asked to speak to me. She burst out crying and told me that she had gone to her thesis professor, one of my colleagues, and asked him how she should go about getting a job, following her Ph.D. He reportedly said, "Jobs are hard to get. Why don't you have your kids first and then we will talk about it." I am ashamed to say that I thought that perhaps she wasn't very good, and that the man was just reluctant to tell her that candidly. I should have known better. At a cocktail party shortly thereafter, I asked him about her. He told me, "She is one of the best students I have ever had." It was as if a bell had rung. I became a feminist on the spot. And I began to interact with a number of wonderful, exciting women faculty in the medical and dental schools. Until 1972 there were no female professors in the university's College of Arts and Sciences, which encompasses most undergraduate and all graduate studies in the humanities, social sciences, and natural sciences.

An aspect of the American Physical Society which had profoundly irritated me over the years was that at its meetings there was a desk labeled "Ladies' Registration" and there was a Ladies' Program. These were directed to the spouses of the registrants, most of whom were indeed women and almost surely ladies. But those of us who were women physicists clearly could not be ladies since we registered with the participants. What's more, why shouldn't our spouses have a chance to go to museums? It may sound a trivial point, but it was very annoying to be made to feel so blatantly that our gender made us marginal. I tried for years to get the programs relabeled "companions' program," but I was unsuccessful until women began to become officers of the American Physical Society, and of the International Union of Pure and Applied Physics, which endorses international conferences. I thought it mildly amusing that much of the opposition appeared to come from the wives of the more elderly physicists, who had organized many of the companion shindigs and who did not want to let go of this social activity.

I was happy to organize the meeting on "Women in Physics" in January 1971. Its participants were Betsy Ancker-Johnson of Boeing (later Assistant Secretary of Commerce for Science and Technology), Allan Bromley of Yale, Henriette Faraggi of Saclay (just elected the president of the French Physical Society), Gloria Lubkin of Physics Today, Enid Sichel of Rutgers, Charles Townes of Berkeley (the codiscoverer of the laser), and C. S. Wu of Columbia. I moderated the meeting.

As soon as the names became public there was trouble. A militant group of anti-establishment scientists promoted confrontations as often as possible at meetings of the American Physical Society. They decided to do it to the Panel on "Women in Physics" under the argument that men, the "enemy," should not participate: it should be an all-female panel. Most of the militants were men but I got in touch with two or three women in the group and had a long talk with them. I told them that since women constituted 3 percent of physicists (now the number is about 9 percent), and had no power within the APS, it was an exercise in futility to just have women speakers. No one would listen to us, and did they really want that to happen? On the other hand, Bromley and Townes were not only very good people, and sensitive to the issue, but they had clout and could help. The women agreed to do what they could, but they were not certain that they could rein in their male comrades. So I got several strong and tough-looking ex-students of mine to guard the microphones and be ready to stop intemperate behavior.

The meeting was a great success. Six hundred members of the APS attended it, and there was a great deal of interesting discussion. An article in the April 1971 issue of "Physics Today" describes it. A committee on the Status of Women in Physics was formed by Vera Kistiakowsky of MIT, and women started becoming officers of the American Physical Society. We became legitimized and visible. However, there are still very few of us who are tenured faculty members of research universities in the United States.

In December 1971 the provost of the University of Pennsylvania gave a talk at the yearly meeting of the American Association for the Advancement of Science in Philadelphia. I went to listen to him. He stated strongly that he and the university wanted able women to become faculty members in the sciences, and he said that the only problem was locating them.

I listened to the provost, and I believed that he meant what he said. Two of my closest women friends at Penn who also heard him did not share this view. Helen Davies and Phoebe Leboy are now professor of microbiology and professor of biochemistry, respectively, at Penn. More than anyone else, they have been responsible for the increasing number of women on the faculty. They are both very, very bright and very determined. The many women now on the faculty of the University of Pennsylvania owe the opportunities that they received to the influence of Phoebe and Helen.

Knowing that the department was about to recommend tenured positions for three white male physicists, I wrote a letter to the chairman, offering my candidacy as professor of physics. I realized that the department would oppose my appointment, but I thought that the provost would back it. I felt that I could demonstrate that I was a more accomplished and better physicist than two of the three people who were to be appointed.

Several of the department members were, in my opinion, outright sexists. Others were no doubt intrinsically opposed to someone who was proposing her own candidacy instead of waiting to be put forward by one of the departmental "barons" or by a subfield fiefdom. There was no chance whatsoever that the nuclear physics group would back my candidacy. And I realized that it would be very difficult for the faculty to accept that the wife of a colleague could be a first-rate scientist. However, I was surprised by the size of the negative vote, and by the intensity of the reaction by the department to my application. Then two marvelous things happened. In response to my request for the reasons why the department had voted against my appointment, the chairman replied that (1) I wasn't sufficiently active in nuclear physics, and (2) I was too old (I was forty-six years old).[12] The second miracle was that I received a phone call from Tom Tombrello of Caltech telling me that the Nominating Committee of the Division of Nuclear Physics of the American Physical Society, composed of himself and Dick Diamond of Lawrence Berkeley Laboratory, and Joe Fowler of Oak Ridge (both members of the Kiev Collective) had nominated me to run for vice-chairman of the division for 1972–73 (and chairman the following year).

I understood that if I won the election, the first reason given by the department against my appointment would sound ridiculous to most people. As for the second reason, it was illegal.

I thought for sure I would lose the election if I ran. The man running "against" me, an excellent physicist, would win hands down. But I was very much involved in the affairs of the DNP and there were things I wanted to do for my field as one of its officers; so I decided I would run: it was an honor to be asked. After the election was over in March 1972, I called the secretary of the DNP from the airport in London, where I was with Wally who had been attending a conference at Oxford, and found out to my amazement that I had won: I was the first female officer of the American Physical Society. I thought, "This does it," but while I became very active within the Division of Nuclear Physics, nothing changed at Penn. The provost was promising action, any month now, but as far as I could see he was doing virtually nothing. There was no grievance mechanism within the university. The provost had appointed an ad hoc committee to look into the matter, but it was essentially willing to continue with the status quo.

Wally was in an impossible situation. Of course he had ceased participating in departmental meetings that might have anything, however far-fetched, to do with me, but the vibrations in the department were terrible. I would enter the building, and while people remained polite, they were very visibly angry. It was extremely depressing. I knew that I had to stay outwardly calm, and not lose my very bad temper, because this would make my life impossible in the department when I won. My physician, at my request, prescribed tranquilizers. I later calculated that I took one kilogram of tranquilizers over the two-year period of the struggle. I also got enormous amounts of psychological help, good advice, and affection from my friends at Penn, Helen Davies and Phoebe Leboy. Wally had received a Guggenheim Fellowship and he decided to go on leave to the CERN laboratory, in Switzerland, during the fall semester. We phoned each other daily and we visited each other several times during the months he was away. I was happy that he was not at Penn.

And then I became very angry. One of my friends in the administration met me and told me, "Fay, you must stop your fight or there will be things that will come out about you that you will regret becoming public." I think that he meant it as a kind warning, but it infuriated me. My life was an open book, and I couldn't imagine what anyone could find with which to blackmail me. In any case, since the days of Mother's red coat in Paris I was not willing to be blackmailed. Later an unguarded com-

ment by one of my colleagues led me to realize that it
was probably my miserable performance at Columbia
which was at issue. But I was neither ashamed of it, nor
had I tried to keep it a secret. I had openly used my
failures for years in advising students.

I was also furious that the provost was unwilling to
act. So I decided to make a formal complaint to the Equal
Employment Opportunity Commission of the federal government and to the Human Relations Commission of the
Commonwealth of Pennsylvania. Affirmative Action requirements are applicable to institutions like Penn, much
of whose funding comes from federal and state taxes
even though it is a private institution. A wonderful professor at the Law School, Howard Lesnick, took my case
on a no-fee basis, as a matter of conscience. However,
we still carefully kept the matter out of the newspapers
because we knew that this would further polarize people.
The question was how could one prove that I was active
in the field. Being elected chairman of the Division of
Nuclear Physics was nice, but could one come up with
something that would be even more convincing to a
group of nonscientists in the government?

I think I was among the first people to think of using
citation counts (another was Robert Davies, Helen's husband). The Institute of Scientific Information keeps track
of the number of times each scientific paper is quoted
(cited) in any other paper. If you are quoted a lot, it
means that your work is exciting to many other scientists, or (as in my case) that it is extremely useful to
many other people, or that it is wrong (and many people
are writing papers to disprove yours). I was very careful
to specify the caveats, but I produced a report which
gave the citation profile for each member of the department, including the newly tenured faculty. One person

had a higher citation count than I, Bob Schrieffer—a Nobel Prize winner. I explained very carefully that this did not mean that I considered myself the second best physicist in the department, but surely I was an *active* physicist! Of course there were additional delays, but finally the federal EEOC, overloaded with cases, transferred my case to the Commonwealth HRC.

The transfer meant that one agency, the Human Relations Commission, could come up with a decision, and that the federal agency would abide by it. In the summer of 1973 the HRC concluded that there was a prima facie case of gender discrimination and stated that the university had until October 1, 1973, to offer me an appointment as professor of physics, with tenure, retroactive to July 1, 1973. If the university did not agree, the Commonwealth would sue the university. It probably did not hurt that my case officer at the HRC, a superb, very bright young black woman, was treated unpleasantly and with condescension by some of the university officials she had interviewed.

The physics department sent out for more letters on me, and it voted again. On October 1, I found out that the appointment had come through, and I became the second woman professor in the School of Arts and Sciences. The first, a distinguished psychologist, had been appointed one year earlier. I calculated that it cost me fifteen hundred very painful hours of my life.

I want to add two things. Once I became a member of the department I felt totally accepted, if not loved. And I have been able to be completely involved in the governance of the university. I am happy to be a professor at the University of Pennsylvania. Parenthetically, I would also mention that men, or at least the men I know, appear to be impressed by toughness. If you win, you're in.

Caltech had been unbelievably supportive of me. At one point, during my Penn fight and its immediate aftermath, the National Science Foundation, my research granting agency, no longer wanted to pay my full-time salary. I understood their view that the federal government should not fully support a senior faculty member at a university. But if I could not receive funding, I would not be able to support my assistant, Catherine Busch, and that was truly disastrous. By then, my review work with Tom Lauritsen required my reading and absorbing over one thousand papers a year in the field of the nuclear spectroscopy of the light nuclei. The papers had to be read and catalogued. Many letters had to be exchanged with their authors. And typically, once a year, we produced a review paper of some two hundred pages in the journal *Nuclear Physics*. The work had to be fast, and very accurate. In 1961 Kit Busch had become my assistant at Haverford and, fortunately, she had followed me to Penn, from which she held a B.A. degree. Kit was extremely intelligent and a perfectionist. She was very well organized, and she was a calm and sensible person. Had my grant stopped, Kit would have had to find another job, and our review work would have been delayed by at least a couple of years. I was trying to think of what to do when I had a call from Tom Lauritsen. The Kellogg group had decided to provide the funds I needed. Out of their National Science Foundation grant they paid Kit's salary, the equivalent of my salary, the necessary supplies, and publication and travel funds. I had not asked for this help. It did not even occur to me that it was possible. I shall never forget that Caltech helped me to continue to work. As for Kit, she retired in 1979 at the age of 68, and she remained my wonderful friend until her sudden death in January 1992.

I also had considerable help from other friends and colleagues. Chien Shiung Wu brought me a rubbing from China depicting three "warrior women." Young friends in the department presented me with a poster which is still in my office and which states: "When you are up to your ass in alligators, it is difficult to remind yourself that your initial objective was to drain the swamp." These and many other expressions of support from colleagues at other institutions sustained me and will always remind me that I should do my best to see to it that an able woman has as many opportunities as does an able man.

Tom Lauritsen was dying of cancer while the Penn fight was going on.

I went to visit Tom and I was overwhelmed by my affection for him, and by his extreme courage and gallantry. I know that I would not have been capable of it. Margie was also very strong and almost serene. In the middle of October 1973 I left to give a series of Sigma Xi lectures in the Midwest. Margie and Wally told me to go, saying I had an obligation since the arrangements had been made long in advance. I called Wally from the Chicago airport to find out if anything new had happened. Yes, he told me, Tom had finally died. I realized that it was the best possible outcome but I wanted to be with Margie. But there was nothing I could do. I gave lectures for several days in Kansas, South Dakota, and Minnesota, and then flew from Minneapolis to Los Angeles.

I stayed with Margie and she arranged the equivalent of an Irish wake. Margie, her children, Willy Fowler, and I ate and drank, and recounted funny stories about Tom.

And, later, she asked me to go through his papers at Kellogg. Margie and Eric Lauritsen (Tom's son) sat with me while I triaged the papers. I asked for a small number of the one thousand pencils I had given him at the beginning of our collaboration. I thought I could calmly go through his files. After all, I had already done it for Misha, in 1962. But it left me drained and unglued. Willy and Tom Tombrello and their Caltech colleagues immediately started arranging a Memorial Service for Tommy. It took place at Caltech on November 1, but I could not attend it. I was at a meeting of the Council of the American Physical Society to ask them to approve a special Memorial Session for Tom, on behalf of the Division of Nuclear Physics. It occurred at the April 1974 meetings of the APS in Washington. During Christmas in 1973, Margie joined us in San Francisco. We stayed in marvelous rooms overlooking the bay, ate at interesting restaurants, and enjoyed San Francisco, and we remembered Tom with love and laughter.

In the late 1970s I became very much involved in science politics.

Nuclear physics was being devastated by what I viewed as an excess of constructionitis. Nuclear physicists are interested in many diverse questions about the nucleus and the forces between the nucleons in the nucleus. Studying the nucleus requires physicists with different abilities and dreams. Some prefer to work individually, or in small groups. Some questions require very large-scale instruments, many scientists, and huge government funding. Nuclear physics is a much more diverse field than is, say, particle physics, where most work requires the highest possible energies. It seemed clear to me that our field should proceed cautiously.

When a very large-scale facility, the Los Alamos

Meson Physics Facility (LAMPF), was proposed in the 1970s, one had to worry not only about the initial capital cost, which could be obtained from Congress by political means, but one had to worry about the continued operating costs. If one assumed that nuclear physics would not get an ever increasing fraction of federal science funds, then every large facility, which costs much more to operate than the smaller ones, particularly those at universities, would contribute to the demise of other areas of nuclear physics. One could not object to that if the large facility would in fact lead to better physics, but I was concerned by the methods used to determine this.

What happened was that the government, which did want the facility because it would strengthen the applied mission of Los Alamos, appointed a committee to determine if the scientific arguments for LAMPF were such that it should be built. They did not load that first committee well: it, in effect, said "no." So, naturally, they appointed a second committee which was more enthusiastic and said "yes," but *only* if the remainder of the field did not suffer as a consequence. I came to understand this type of recommendation well. Yes, the machine would be built, but the "environmental" statement was sure to be neglected. I wanted the community of scientists to have a way of presenting their views directly to the agencies, rather than through committees controlled by the granting agencies. Without going into details, I made a nuisance of myself for about a dozen years on these and related matters. As a result, the community became better informed about the changes as they were occurring, but as soon as one of my angry young colleagues was co-opted into an official committee, his views, with rare exceptions, became those of the establishment.

I had very little influence as a member of the Nuclear

Science Advisory Committee, appointed by the Department of Energy and the National Science Foundation, in advising the agencies about funding scenarios. I was able, however, to make its members aware of potential conflicts of interest, and to deal with them cautiously. While I was on NSAC, Willy Fowler and then Herman Feshbach of MIT were chairmen of the committee. I like Herman tremendously. He is an excellent physicist, and an extremely bright and honest person, with a sharp wit. We fought openly during the meetings, and were good friends at other times. Our robust discussions in these and other physics committee meetings would end with the chairman saying, "Well, are you guys all agreed?" And while I often disagreed, I was glad to be one of the guys.

On the whole I think that I was correct in my concerns about the development of nuclear physics. It would surely not have been in the interest of science to have nuclear physics remain an artisanal field. However, I still think that some of the large facilities should not have been built: the resulting scientific contributions have been modest, but their costs have skewed the field and may have contributed to its relative decline. I do regret several of the strong and tactless remarks I made at open meetings.

Of course these activities took a lot of time, and I was frequently out of town because I was using facilities at Los Alamos (not at LAMPF) to continue my experimental research.

I worked for seven years with Ed Flynn at P-9, the bureaucratic name of the Van de Graaff facility. The facility had a superb machine, excellent staff, and a fantastic view of the Sangre de Christo range to the east. I worked there for a few days during school holidays, during the summer months, and on a sabbatical leave. It was a very happy time because I loved working with Ed and my other colleagues on the Hill. Ed Flynn is a tall, lanky Midwesterner who received his Ph.D. for work done with Louis Rosen, a member of the Kiev Collective, who became the founder and the first director of LAMPF. Ed is extremely smart and he is a warm and charming person. He had a very happy marriage for some twenty years and had six children. Then his wife had a stroke during a snowstorm, and by the time he got her to a hospital she was effectively brain dead. He did everything he could for her for another year and a half before she died. I mention this because during that time he tried very hard to understand whether part of her brain might still be working. Finally he left nuclear physics, and started measuring the patterns of the magnetic fields originated by currents in the brain, when the brain is stimulated. His ingenious work is world-class and has led to a new understanding of the way in which the brain works. He showed enormous courage in switching fields in middle age, and the very high ability to succeed. Ed is now married to Maureen Flynn, a delightful woman, and he has an army of grandchildren.

I would fly from Philadelphia to Albuquerque and rent a car at the airport. Then I would switch the radio to one of the Spanish-language stations and drive north to Santa Fe. By the time I reached its outskirts I was starved for oxygen (both Santa Fe and Los Alamos are at about 7,500 feet) and shaking from the strain of driving an unaccustomed 75 mph. (People drive very fast in New Mexico and it isn't safe to be overtaken constantly by trucks.) Sante Fe lies in a high basin, surrounded by a beige and brown moonscape, underlined by tufts of green bushes and, in the fall, by the deep yellow of cottonwood. The vertical layers of the mountain ranges are often separated by shimmering veils of fog.

When I reached Santa Fe I would sigh with relief, take a quick look at the city, stop by my favorite grocery store, Kaune's, and drive north and then west to Los Alamos. The last part of the drive, from the Rio Grande Valley to the top of the mesa, along a winding road, was my favorite. On the mesa I stayed at the Los Alamos Inn and, for longer stays, in small, rented apartments. Los Alamos has never lost its magic for me. The contrails in the sky are more beautiful and the afternoon thunderstorms bigger and better than anywhere else. The people there seem more real and more relaxed than in what my chums call the "effete" east. And the arts of the region are superb. It is also the only place where I have enjoyed opera. The stage at Santa Fe is open to views of the mesas and evening lightning storms.

And the work we did was very interesting to me. We used a magnetic spectrometer with a very fancy detector designed by Ed. We would run for several days at a time, twenty-four hours a day. Even when I was not on duty, it was hard for me to sleep because I wanted to know what was happening. I doubt that any drug-induced high can match the exhilaration I felt as new peaks appeared on my computer screen, responding to the signals generated in the "target room" where the spectrometer was located. Because of health hazards, the target room and the room where we, the experimenters, visualized the results on computer screens had to be separated by about 500 feet. One of the hazards arose from the particles—tritons—which we accelerated in the Van de Graaff. Tritons are the nuclei of the heaviest type of hydrogen found in nature. Tritium is radioactive, and is a particular hazard because it can replace hydrogen in human tissue. As a result, very few places in the world work with tritium, although a number of interesting nuclear phenomena can best be studied with tritons.

Throughout the days and nights we literally ran between the target and the computer rooms, typically every hour or so, changing the geometry of the equipment, the parameters of the detectors, and replacing the targets of the nuclei we were studying, as they broke. As is typical in all of experimental science, equipment never worked the way it should, particularly on weekends and at night when it was most difficult to request help from the excellent technicians. These were among the earliest experiments in which a computer was used "on-line," that is, while the run was in progress. Fortunately one member of the group, Jules Sunier, was both expert and very nice. During one very long night, when I was the only experimenter on duty, he had the computer beep at me (which got my attention promptly because I assumed that something had gone badly wrong) to give me a message, in French, that he would soon relieve me, and not to despair. I could last for about thirty hours but then I would shake with tiredness and, I realized even at the time, with bliss.

In addition to the research at Los Alamos, I was also very involved with professional organizations.

I had been elected a member of the Council of the American Association for the Advancement of Science, and appointed first a member, and then the chairman, of the Commission on Nuclear Physics of the International Union of Pure and Applied Physics. At that time IUPAP had a very important role to play in validating international meetings. The Committee had twelve members from different countries. We received requests from meeting organizers to have their meeting be IUPAP-sponsored. What this meant is that we decided whether the meeting was well organized and whether it responded to a genuine need in the field. We tried to make sure that there weren't overlapping meetings.

Funds to travel to a meeting were more easily available to scientists in all countries if they could state that the meeting they wanted to attend was IUPAP-sponsored. But the most important part of the approval was that it required the meeting organizers, and their government, to pledge that they would not deny access to their meeting to scientists from a particular country that might be on their political shit list. Exclusion from an IUPAP-sponsored meeting could not be due only to the scientist's national origin. It meant, for instance, that I had to threaten not to give approval to meetings held in countries which denied access to South Africans, or to Israelis, or to Taiwanese, or whatever. This policy helped to make meetings truly international. We decided most issues by mail, but occasionally the Commission would meet—in Berkeley, Versailles, Dubna, Munich, and Tokyo during my tenure.

The Commission meetings were held at the times of regular international conferences. The one at Dubna, in 1976, also involved one of my more exciting trips. Dubna is the site of the Joint Institute for Nuclear Research and is located about seventy miles north of Moscow. The basic cost of the trip (direct economy air fare, and room and board at the conference site) was paid out of my research grant. I decided to embroider it with my own funds. I bought a EurailPass and flew to Paris. There I took an overnight train on the long trip to Vienna. When I arrived tired, hot, and grumpy, I took a cab to the Hotel Imperial, where I found a large bouquet of flowers from Wally which revived me. The next day I flew to Moscow and the following day I was driven to Dubna. Dubna was particularly beautiful and the lilacs were in bloom. One of my favorite friends there, G. N. Flerov, had me to his house. He knew I like caviar and I have never eaten as much. I had gotten to know him well at a meeting in Japan when he asked me to be his translator. Flerov died two years ago. He was one of the people instrumental in getting the Soviet Union to embark on the atom bomb program. He was reputed to be a very dominant presence in his lab, but I always found him a charming and charismatic person. His institute has now been renamed the Flerov Institute for Nuclear Reactions, and its director is another friend of mine, Yuri Oganessian.

After Dubna I took the overnight train to Leningrad, spent a couple of days sightseeing, and took a plane to Helsinki and then on to Turku. I wanted to see the Aland Islands of my parents' youth, and I traveled on a ferry to Stockholm. The weight of my baggage had increased greatly due to the many gifts I had received in Dubna, including heavy vinyl records of most of Tchaikovsky's works, and art books, and there were no baggage handlers anywhere from Turku to Copenhagen. Then after a wonderful visit with the Bohrs, I flew home from London. It was the last time I saw Marietta outside a hospital. She died of cancer in 1978.

Wally was working at Fermilab, near Chicago, much of this time, and he too was traveling to conferences all over the world. In the 1970s, after the serious problem with the work at Cornell, he rebuilt his research group, and organized and led a collaboration which carried out a major program at Fermilab. He benefited immensely from the work of his original and creative research associate, Walter Kononenko, with whom he has collaborated for twenty-five years. In 1978 Wally gave invited papers on this work in France, at Copenhagen, and in Kyoto, and I accompanied him on those trips. But we were apart a great deal of the time.

Our telephone bills over the years have been staggering. Except when one of us is in the USSR, we call each other at least a couple of times and sometimes as many as a dozen times a day. It was often ridiculously difficult to see each other. We would make a date to meet in airports, and soon knew the best trysting places within hailing distance of the gates. Our view has always been that it is wise to trade money for more time together. Once I had been to a meeting at Aix-en-Provence and he was going to CERN, in Geneva. I took an overnight train from Marseille to Geneva. He met it and we had a couple of days together. Another time I had a meeting in Osaka. He flew to Honolulu. We met there and flew on to Kona, for a vacation. And there was the mad time when I had been working at Daresbury, near Liverpool in England, for several weeks, and was going on to a meeting in Visby, on the Island of Gotland, in the Baltic. He flew to Heathrow. I drove there to pick him up, we went to Bournemouth for a weekend, and then back to Heathrow. I flew east to Stockholm and he returned home. It has not been a conventional marriage, but it has been great.

Of course, this enormous freedom was the result of our being childless. As I mentioned earlier I was unable to conceive. We tried all possible research treatments during the first ten years of our marriage, at least one of which led to continuing medical problems for me. Over the years it has become a faint, residual regret for both of us. Still we have the joy of working with students and watching them develop, and it has been a very close and a very happy marriage. Our housekeeping chores are easily met. I cook the evening meal. Wally prepares breakfast and a brown-bag lunch because I am dysfunctional in the early morning. When we have guests I plan and prepare the food, and Wally washes up. When either one of us is out of town or is ill, the other takes over. We shop together in the pleasant stores of this area and at the Wayne Farmer's Market ten miles away. Wally is in charge of dealing with repairs and the upkeep of the house because we have learned that service people and tradesmen will disregard a plea to come "first thing in the morning" if a woman makes it. They more readily understand that a man has a job that he must go to! We even manage to live comfortably together in a house part of which is messy (Wally's office has spilled into the dining room), and part pristine (mine—I cannot stand disorder where I work). Above all, Edith Briggs, our housekeeper and friend, sees to it that it is spotless and comfortable.

I owe Helen Davies and Phoebe Leboy two things: my job, for which they had set the foundation through their activism; and surviving the two years of the struggle over my appointment as professor of physics at Penn.

So when my appointment became official, I asked them what I could do, in turn. They laughed and said that it was nice to have another woman to serve on committees: they had been token women on too many of them. I therefore found it very difficult to decline any requests to serve on Penn committees. Some were pretty trivial. Others were more interesting. The committee structure at a research university is very important. At Penn, the faculty maintains a remarkable amount of self-governance, and does so through a series of both permanent and ad hoc committees.

I was appointed the chair of a committee to recommend whether one of the smaller schools of the university, the School of Public and Urban Policy, should be phased out. I smelled trouble, because the president of Penn was a faculty member of the school at the time. After several months of very lengthy and thoughtful deliberations, my committee concluded that SPUP should be phased out. The president exploded and appointed another committee. Several months later the new committee came to the same conclusion. SPUP no longer exists. What still does exist is my friendship with Morris Mendelson, emeritus professor of finance in the Wharton School at Penn, with whom I served on that committee and on many since. He is a bright, warm, and delightful man, with a puckish sense of humor, and I have learned a great deal about the university from him. Morris is also a gourmet cook and his recipe for roasted fish is superb.

There was also the committee to choose the next president. I was one of four faculty members, and joined two students and six members of the Board of Trustees in seven months of deliberations and interviews. With one exception, I was enormously impressed by the trustees. They were men and women of exceptional intelligence and savvy, and worked for Penn with great dedication. The one exception was a trustee who was a faculty member at another institution. He unfortunately had great prestige with the other board members. I found him arrogant and opinionated. His comments effectively blackballed several excellent candidates. The trustees had a different view of life than the one I was accustomed to. Once four of us flew south to interview a candidate. We had to change planes in Washington but found the connecting flight was canceled. Paul Miller, the chairman of the board, was unfazed. He said, "Just a minute," hired a plane, and we took off. I remember the flight well. I was sitting next to Gloria Chisum, a very strong and able woman, who, among many other activities, designed cockpits for the Navy. (She is now vice-chairman of the Board of Trustees of Penn.) The pilot addressed her and me in a condescending way, and explained the gauges to us. We raised our eyes heavenward. To keep our meetings with candidates away from Penn's gossip trails, a number of them were held in corporate board rooms in Philadelphia. I learned also, for the first time, the meaning of the word "feisty," which I had never heard before. I asked Wally what it meant, after one of the trustees said, somewhat disapprovingly, that I was feisty. On balance, I agreed with her.

While it might not be apparent that I had time to do so between 1974 and 1984, I did teach full time at Penn, and enjoyed it tremendously. I had been made to feel welcome by Walter Wales, who became chairman of the department in 1973, and remained chairman for nine years. Walter is a man of integrity and savvy, and he is a superb teacher. He has the most common sense and wisdom of anyone I know, and I respect him tremendously. He has an excellent sense of humor, and he doesn't take himself too seriously, a trait very rare among faculty. Walter helped make my life at Penn comfortable and interesting. He is now deputy provost of the university.

In September 1982 I went to see Jack Mackie for my regular checkup, and the mammogram showed something unusual in my left breast.

Another mammogram a couple of months later also looked suspicious. I became very worried: I had expected the first breast cancer, but I had caught it early, and twelve years later I was not ready for further trouble. I tried to concentrate on my teaching and I went to a conference at Michigan State University. I had a good time, and it was fun to see several of my Russian colleagues again. We were having a drink together when one of them asked me what I thought of President Reagan. I said that I disliked his domestic program, but that I agreed with his views on foreign policy. They burst out laughing. But I was beginning to feel pretty grim. I went to see Rosalind Troupin, the radiologist who had been checking me, and I asked her what she would do if the mammogram were hers. She was wonderfully candid: she said she would ask Jack to do a biopsy and proceed to surgery on the spot, if necessary. I waited until the end of the semester and went to the hospital in December.

Jack decided that I would first go via Rosalind's lab where she would locate the anomalies with the X-ray machine, insert needles to show where he should cut, and then I would proceed to the operating room. Jack was as concerned as I was. When the attendant who was supposed to wheel me down from my room failed to appear, he cursed and pushed the gurney himself to Rosalind's lab and then to the operating room. I was anesthetized. When I woke up, Wally and Jack told me that it was indeed cancer. A week later Jack came to me looking stricken and told me that this second cancer was different from the first, that it was dispersed throughout the ducts of the left breast, and that one of my lymph nodes had been invaded. He told me I probably needed to begin chemotherapy, and he induced a top-notch oncologist to take me on.

John Glick is now the director of the Cancer Center at HUP (the Hospital of the University of Pennsylvania). He was then only in his thirties but already had a

national reputation. He was clearly superbly intelligent
and totally savvy. Wally attempted to obtain quantita-
tive information about the likelihood that chemotherapy
would improve my chances to survive. John was candid.
Very little information was available for patients of my
age who, like me, were premenopausal and had the
wrong kind of estrogen receptors. On balance, he said,
he would recommend chemotherapy, starting immediately. I refused. Wally and I needed a break. We had
been planning to go to London for the New Year vacation, as we often do, and I insisted that we go. I was still
in severe pain and not very mobile, but we flew the
Concorde to London, and I was taken in a wheelchair
through Heathrow. It was a wonderful break. We stayed
at our favorite hotel, in our favorite room, and we had
lots of tea and cucumber and smoked salmon sandwiches.
We walked around London, bought piles of books and a
new cashmere sweater for Wally. We communed with
the ducks in St. James Park, saw a couple of plays, and,
even, toward the end of our stay, went swimming in the
hotel's pool, which has a magnificent view of London. It
was then that I began to realize that the nature of the
scars left by the two mastectomies was such that I would
have residual discomfort for the rest of my life. I always
feel that a too-tight band is binding my chest.

Then, of course, reality caught up with us. I started
teaching the second semester, and I began the chemotherapy. I was soon feeling very, very ill. The chemo
came in cycles: injections, then pills by mouth, then a
short period of peace, and then the cycle began anew.
The injections were miserable. I have a long-standing
phobia about needles. My veins are hard to find and roll
out of the way of the needle. I imagine all too clearly a
needle being moved under my skin to find the vein and

penetrate it. The extremely nice and skilled nurses in John's office allowed me to lie down while they injected me with that day's poisons. It was not pleasant. And then, very quickly, as I had been told I would, I lost my hair. I thought it wouldn't matter but it did.

I am short and heavy. I always felt that only my coarse, but full and wavy hair is beautiful. I knew it would come back, but in the meantime I had to buy a couple of wigs. At both stores the saleswomen were unpleasant. I felt helpless and manipulated, and I felt very different wearing a wig over my totally bald head. I was unconcerned about the loss of my second breast. The prosthesis I had worn for a dozen years was heavy and very hot during the summer. I experimented with loading two prostheses into my brassiere and then hauling it against my chest. I tried stuffing some padding into the brassiere. Finally, with Wally's approval, I decided not to wear anything. I try to dress in loose tops as often as possible since most women are very scared about losing their breasts, and I do not want to remind them of their fears.

I was lucky that I did not vomit often during the chemotherapy. An antinausea drug which works for some one-quarter of patients worked for me. But I did manage to have several of the side effects that only one percent of patients develop. Some were trivial, others were not. All were characterized by frustration at John Glick's responses, which were variations of "You are dreaming up this problem," "OK. It is a real effect but it is due to your becoming older," and "No, there is no way that it can be due to the treatment." John insisted that all side effects would disappear within a year of the end of chemotherapy. But, in fact, there have been lingering problems. One of them is "dry eyes," as a result of which my eyes are always sore. John said "Yes, I did it to you" three years later, after dismissing all my earlier comments. I accept the fact that when one attempts to combat cancer by means of a therapy that is not well understood, one cannot know all of its effects on an individual. I blame no one for the bladder cancer I developed five years after the chemo, almost surely because of the cytoxan I was given. I accept the fact that to a physician the primary problem is fighting the disease, and that other problems are subsidiary. I found it extremely difficult, however, to be told that I was dreaming up symptoms and that they were unrelated to the therapy. In a sense it was a criticism of my objective powers of observation, which are an intrinsic part of my being a scientist.

The worst side effect at the time by far was that which affected my mind. John had decided that I should take a drug to help me over the depression caused by the chemo. One common side effect is dry mouth, a decrease in saliva in the mouth. There is, however, a less common but devastating side effect: severe memory loss—and I developed it. It was an unnerving experience to stand at a blackboard and grope for the word "axis" and be unable to think of it. And the more often this happened, the more anxious I became. John insisted that the drug could not have anything to do with my memory loss, and that I was just growing older. As a scientist I did not believe that a sudden memory loss, occurring at the same time as a new drug was being taken, could be due to the more gradual effect of becoming older. As a person, however, and particularly since John appeared to be so certain that there was no connection with the drug, I was extremely worried that I was suffering an irreversible memory loss. My hair would grow back, but my mind was becoming less sharp. I had to stop teaching that semester though I resumed teaching in the fall. After a year elapsed, I insisted on discontinuing the drug and quite shortly thereafter I began to be able to think clearly once again.

John is clearly brilliant, but at the time he seemed to me to be insensitive. For a couple of years, whenever he saw me during my checkups he would comment on my hair ("The wig looked better"), my clothes ("frumpy"), and my shoes ("awfully ugly"). He may have thought the remarks were kidding, but they were repeated each time I saw him. Wally, who regularly went with me on my visits to John, found them very offensive, but he advised me to continue to see John to benefit from his great expertise. After some time, Wally suggested that I go to my next checkup alone, and speak to John about his remarks. He stopped making them. I think that altogether he has become a more mature human being. John wisely suggested that we do one "fun" thing a week together: we call them "Glicks," and we go to the IKEA store to buy my favorite Swedish herring, or to a movie 20 miles away, or to a new, and possibly pleasant restaurant. Wally sustains my long-range interest through major "Glicks." In a memorable MegaGlick we flew around the world during the Christmas–New Year recess in 1991 via Hong Kong, whose beautiful harbor was lined with gaudy lighted-up skyscrapers, in a celebration of Xmas run amuck. Then we flew in a great arc to London, over India, the Himalayas, Moscow, Riga, and Hamburg, while watching four movies on the VCRs available at each seat. Margie Lauritsen, who married Bob Leighton of Caltech several years after Tom's death, met us in London and then we all went to Arosa, Switzerland. We celebrated New Year's Day with Denys and Helen Wilkinson in Sussex.

Jack Mackie also operated on Wally in July 1983. It was a quite major intestinal problem, and followed an operation for a similar condition in 1971. The problem was not made easier by Wally's insistence that yes, he could and would work in the hospital, which required me and several of his young friends to trundle a PC and a printer to his tiny hospital room. (Today such equipment would be both much smaller and much more powerful.) The equipment and Wally got in the way of the nurses, and I was told that he was put on their "undesirable" list. Nevertheless, he recovered and in late August we decided to go to Europe. The main international conference in my field, which occurs every three years, was taking place in Florence, and Wally accompanied me. We flew to Copenhagen, had a brief vacation in Switzerland, and then took a train to Florence. During the days when I was at the conference, he rested in our room, which had a magnificent terrace overlooking that beautiful city. It was too hot to wear a wig. I now had a downy stubble on my head, and looked like a plucked chicken, and Wally looked pale and wan, but our friends were quite wonderful. I thoroughly enjoyed the conference and a party one evening, at the house of Italian friends in Fiesole, with a breathtaking view of Tuscany. We were driven to the Pisa airport, following a quick detour to see the Leaning Tower, and flew on to Paris. This was a cossetted and pleasant trip. It did not really show me whether I could survive on my own under difficult conditions.

What really improved my morale and contributed to renewing my joie-de-vivre was a trip to Alma-Ata in Kazakhstan in 1984. Soviet scientists in my field have a yearly meeting on topics in nuclear spectroscopy. Usually two or three Westerners are invited to give papers at those meetings, whose venue changes from year to year, except that every five years the meeting has been in Leningrad, now again St. Petersburg, where the field had its beginnings in the USSR. I had been invited to the 1983 meeting in Moscow, but I had to decline because of the chemotherapy. I was invited again the following year. I accepted and it was a splendid trip. I finished my classes one Friday in April and flew overnight to London. The following morning I flew to Moscow. I had arranged for a room there because the flight to Alma-Ata was due to leave the next morning at 2 A.M., from Domodedovo Airport. It will surprise no one familiar with the USSR that when I went to negotiate with Intourist for my transfer to my Moscow hotel, the agent said, "Oh, come on, you don't need a hotel. We will just transfer you directly to Domodedovo." No way was I going to sit for ten hours in a miserable airport lounge. So I sneakily used my Russian trump card. I said, "I am a cancer patient," and displayed a letter from John Glick. I regret to say that a statement about *rak* (cancer) is extremely effective. The health system in Russia is so bad that cancer is equated with death, and people become very kind and helpful.

So I was transferred to the National, my favorite hotel, next to Red Square. The National is very old, and both splendid and shabby. (It is now, unfortunately, being renovated.) It reeks of history, and the food isn't bad either—if you can get served. I was given a grand suite overlooking Red Square and had an excellent rest before resuming my trip. Domodedovo was as appalling as I had imagined it to be, and the flight was five hours long and very tiring. We did get the standard Intourist meal of that era: quite tasty boiled chicken, cooked on the plane. In the morning, Alma-Ata and the beautiful Tien Shan range which separates it from China appeared before us. I was taken to a newish hotel where I was given a suite, and after some sleep I went for a short walk. The next morning the meetings began. I gave my talk, which was well received, and I was taken out for drives around the city and in the mountains, and to dinners with colleagues and with local students. I met many people for the first time: only certain privileged physicists were allowed to attend conferences outside the Soviet Union. The younger scientists were particularly delightful, although we could not interact as much as we would have liked because the Soviet scientists were excluded from the hotel in which we, the foreigners, were staying.

The day after I arrived, the manager of the hotel said I would have to move out of the suite since my prepaid arrangements were for a single room. I said that I would be happy to pay the difference, and how much was it? She said, fine, I could stay, and she would get in touch with Moscow to find out. When I was ready to leave I asked her what I owed; she sighed, told me that Moscow had never answered, and to forget it.

In Alma-Ata I saw Claude Detraz, whom I had known since he was a postdoc at Lawrence Berkeley Laboratory, and met his Russian wife, Irina, for whom I developed much affection. When I see them, every two or three years, we immediately resume our friendship. Claude now runs nuclear and particle physics in France, and Irina is a world-class interpreter in several languages at important international meetings and conferences. Her initial training was as a physicist. On the way home from Alma-Ata I spent the five hours on the plane learning a great deal about life and Russia from a very nice woman named Ada Azo who was one of the principal scientific translators in the USSR. In Moscow I once again spent a night at the National and then returned home via London. I had spent nine quite exhausting days on a trip to the other end of the world, and yet I felt energetic and bushy tailed, and ready to return to my class which Wally had taught for the week. I am grateful to the people who invited me to Alma-Ata. The trip showed me that I was back to normal and ready for practically anything.

6

The Anteroom

1984–1992 • • • • • • • • • • • • •

WALLY AND I EACH HAVE TWO OFFICES in our house on Cherry Lane. My downstairs den is paneled and has bookcases, cupboards, files, and an elegant desk that Misha bought me forty years ago. I use it primarily as a library and storeroom. The room where I work and dream is on the second floor. It is adjacent to the bedroom, and when Wally is out of town I isolate the two rooms from the rest of the house. We call this the anteroom, the ante. It is full of books and papers, dominated by Indian and Eskimo art and by a large, comfortable couch. My connection to the outside world is through a small TV, which I use mainly to watch the news, and through a tape recorder on which I play Russian songs, a rather unlikely equivalent to a security blanket. The songs I like best are by Vladimir Vysotsky, a famous Russian bard who died some ten years ago from drink and drugs, and who sang powerfully of misery and of friendship; and also those by Aleksandr Galich, which are more intellectual and political. I also stare at an Inuit carving of a hunter about to stick an ivory knife into a large, dying bear. Except for Vysotsky, the room is very quiet. It is insulated from the street by two other rooms. It overlooks our lawn and the Chinese maple and the two lilacs we have planted. In October the leaves of the maple have become red once again, and birds dart from branch to branch, in a swift, incomprehensible game of tic-tac-toe. The ante is my haven.

It is a good place to think, but as we come closer to

the present, my recollections are no longer seamless and tempered by time. My thoughts come under discrete headings.

1. My Research Comes to an End

After Ed Flynn switched fields and P-9 was no longer used for basic research at Los Alamos, I tried to continue my research on triton-induced nuclear reactions at Daresbury, in England. The people at Daresbury were exceptionally nice, and I loved working in England, but the equipment for the particular work in which I was interested never did work. I enjoyed becoming, for brief periods, part of a British community, but there weren't enough local resources to make the program viable. Once, one of the technicians who was, as usual, repairing the accelerator, which is twelve stories high, said to me, "Luv, come on, I'll show you a beautiful view." It was the Aiguille du Midi revisited. I climbed catwalks behind him, and indeed the view from the top, of Warrington and the countryside, was splendid. I have blocked off from my mind the climb back to ground.

Then I tried to work at the Indiana University Cyclotron Facility in Bloomington. I have many friends there, some from graduate student days, and we began to study light-ion-induced reactions together. Here the equipment worked superbly, but Nature was not on my side. I decided to use a carbon-14 foil as a target, and colleagues at Michigan State University were kind enough to loan me one. Carbon-14 is radioactive and has to be handled carefully, and we did. Finally, we were ready for the experiment. The target was gently placed in a vacuum chamber, the background runs were made, and we were ready to go. It was, naturally, in the middle of a very stormy night. I typed the word "RUN" into the computer, meaning that the computer should start to record the run, when all signals from the carbon-14 vanished. It took us some time to realize that the target happened to disintegrate when I told the computer to start recording.

It was a bad problem. The thunderstorm had gotten worse and all phone lines were down. When radioactive contamination occurs, one immediately has to call the radiation hazards expert attached to the lab. It was impossible to do so. We thought of driving to his house, but it was far away in the hills, and the cyclotron crew advised against it. My chums and I decided to go in the vault which contained the vacuum chamber to understand the extent of the radiation contamination. We went in very, very carefully, and we determined, with radiation monitors, that it was confined to the inside of the chamber. It was thus not a danger to us and to the crew. Because of my classes I had to return to Philadelphia before the spill inside the chamber could be cleaned up. It took the IUCF radiation staff three weeks to do it. I felt very guilty but it had not been anyone's fault. It just happened. My colleagues at MSU who had loaned me the target were extremely understanding.

I returned to IUCF on several other occasions, but I finally decided that it was unreasonable for me to try to work there as a user. I could only come two or three times a year. I came alone. The equipment was sophisticated and the computer software needed to operate it was constantly being improved. From one trip to the next I had to learn anew how to use it. Each time one or two staff members patiently taught me the new procedures. It was unfair to accept their constant help. In addition, I was getting older. Taking the US Air plane to Indianapolis, driving fifty miles in a rental car, staying in noisy motels and eating fast foods, and then undergoing thirty-hour runs on the machine was too much for me. I hated to admit it because I was enjoying myself so, but I couldn't really go on. When I stopped experimental research in 1989 my life became more circumscribed, but I still worked on my reviews and traveled a great deal.

2. And So Do the Review Articles

Gradually I began to think of ending my work on the Energy Levels review articles, which I had started with Tom Lauritsen in 1952. My decision was helped by my repeated clashes with my grant officer at the Department of Energy, to which my grant had been moved some ten years earlier. I probably had the smallest grant in his section, and surely the most cost-effective one, but every year he tried to cut me down. Finally one year he told me that he would cut me below the least amount with which my review work could continue. He told me no one could change his mind. I exploded. I felt that I could not appeal to his superior because he is my friend, David Hendrie. Instead I wrote to a dozen people in the field who used my work and who had clout. They communicated their very strong feelings to the Department of Energy and my grant officer was told to give me adequate funding. But at some point he had asked me a good question: "How much longer do you plan to work on the reviews?" It was clear to me that they were still very useful, but my health problems warned me that I should transfer the work responsibly while I could still do so. I decided to terminate my work in December 1990, and I am happy that a couple of good, and younger, physicists were willing to continue it. When my final paper was about to be published,[13] I found to my amazement that Denys Wilkinson had written a foreword to it, in which he spoke of my work with great kindness and with his usual elegance. His comments have meant a great deal to me.

3. There Have Been Problems

Two other things had happened which were turning points in my decision to stop my scholarly work. The first was a head-on crash in 1987.

I was returning home from Penn and I was driving 20–25 mph on a winding street five minutes from Cherry Lane, when a motorcyclist who had been driving 40–50 mph in the opposite direction lost control going around a curve and crashed head-on into me. I saw him as a blur, and stopped on a dime. His helmeted head smashed my windshield and I became covered with glass. The police came quickly. There was no question whatsoever about what happened since the motorcycle had left long skid marks and my car none; and there were witnesses. He was taken to the hospital, and I was taken home, very shaken and with a pain in my right knee which lasted several months. The police officer told me that if the guy had hit a couple of inches to the left, I would probably have been killed. My car was totaled and it was hard for me to drive for two years.

The eighteen-year-old motorcyclist was slapped on the wrist by having his license revoked for a while. He decided to appeal. The police called me as a witness. The boy's father thanked me publicly for having driven so slowly because otherwise his son would be dead, and the judge upheld the wrist slap. I was very angry because I thought that the guy should have accepted his responsibility, and because he had not apologized for the trouble, and the pain, his carelessness had caused me. So I threatened to sue his insurance company, mainly in the hope that he would have a difficult time obtaining insurance in the future, and the company settled with me. Wally was wonderful. He was willing to help me spend a considerable amount of time preparing the case and talking with our lawyer because he knew that I would not get over the trauma until I could take some action.

The second problem occurred in 1988. I had severe bleeding episodes during the spring. For several months all tests were negative. Then I had a cystoscopy, and a small, malignant tumor was found in my bladder and removed. Bladder cancer is not a common type of cancer for women. It is not life-threatening if treated soon enough, as mine was, but it recurs. It is almost certainly a by-product of the chemotherapy, but there was no way to know that ahead of time. It was an unpleasant shock. At first every three months, and now every six months, I have a cystoscopy. The next one is two weeks away, and I dread it. I know that sooner or later another operation, and more procedures, are likely to take place.

The shock of the second breast cancer, the chemo-therapy, the crash, and the bladder cancer, and my concerns about Wally's health, have all contributed to my having a high level of anger and anxiety. At my doctor's suggestion, I began to take small amounts of an anti-anxiety medication. I am addicted to it but the amount of the drug that I need to take is considered minimal. I constantly strive to use less. I do not like being dependent on a chemical but it enables me to function well and, most of the time, joyously; and it does not appear to affect my mind.

4. And There Have Been Pleasures

In 1987 I went to the University of Sussex to speak at a conference in honor of Denys Wilkinson's sixty-fifth birthday. We first went to a country manor in Devon that we like very much, Gidleigh Park, where I played croquet with Wally. He was very displeased at my playing French-style, which meant that when my ball came to rest against his, I used my mallet to cast his ball into the weeds. He thought it unsporting and refused to play further. Then in 1988 I was asked to lecture at a Summer School in Nuclear Physics in Mikolajki in Poland, 150 miles or so north of Warsaw. I accepted before the bladder operation took place. I should have canceled my trip but I was reluctant to do so a month before the school started. The trip to Mikolajki was all right but the accommodations there were primitive.

The school was held in a worker's hostel. I was grateful that I was given one of the rare single rooms but, of course, it did not have private facilities, just a wash basin. The room was filthy, and the first evening I killed some thirty mosquitoes. I learned to keep the window closed, but it was very, very hot. And the flies continued to skim over my head like buzzards. Once again my colleagues were wonderfully kind, but it was a pretty difficult time. Because of the operation, I needed to void frequently, including at night. I would open the door to my room as quietly as possible and furtively go to the toilet at the end of the long corridor. Then I would have to open and close the door and flush the noisy toilet which probably woke up everyone. Then I returned to the room, and the travail to use the toilet made me want to go again. I could not really go on excursions or sit on the pleasant lawn because the Masurian Lakes district is swamped with mosquitoes and I am allergic to their bites. I remember very little of the details of my stay, except my astonishment at the pollution of the beautiful lakes and my concern for my Eastern colleagues. All the stores were empty and there was very little food. Through my friends' kindness and savvy I was able to talk to Wally by phone a couple of times. He called me to let me know that his brother, Joe, had died and he made reservations for me at our London hotel for the night I was to fly from Warsaw. I was leaving before the end of the school's session because I had to return to Penn to teach. One of my Polish colleagues accompanied me to the Warsaw airport, after taking me to meet his wife. They gave me lunch at their apartment; and he showed me the site of the Warsaw Ghetto.

When I got off the plane in London, and shlepped my suitcase through customs, I found that the hotel had sent a chauffeured Jaguar to meet me. I wonder what the chauffeur thought. I was exhausted, I was dressed in dirty clothes, and I stank because I doused myself daily in antimosquito repellent and the drains of the showers at Mikolajki had not worked and were filthy. When we got to the hotel, one of the assistant managers said, "Sorry, Madam," and explained that they were so fully booked, because of an Air Show, that they had to give me a suite. I contemplated my ill fortune calmly. The bathroom had two robes. I threw all of my clothes in a heap, put on a robe, and called Wally. When we finished talking, I junked the robe and soaked through three baths, one after the other. Finally I put on the second robe and ordered bisque de homard, smoked salmon sandwiches, profiterolles au chocolat, and a large bottle of mineral water.

My final journey *aux frais de la Princesse* (at government expense) was to Kiev in May 1990, to a school sponsored by the Ukrainian Academy of Sciences. Wally accompanied me because we have become increasingly reluctant to be apart. When he completed making the plane reservations on the phone, the American Airlines clerk asked, "Do you mind telling me whether you are going there on business?" Wally said, no, he was accompanying me, and that I was going to be giving some lectures. She asked, "On what?" Wally said, "On nuclear physics." There was a long pause, and then she said, "Well, have a good time." We flew to Kiev via Zurich and Vienna. We were met and driven to a sanatorium some 7 miles south of Kiev. It was a green and pleasant area, and we were assigned a couple of rooms with private facilities. (I had explained that I had medical problems.) It turned out, however, that the school was starting a day later than we had been told and there was no food, none at all. Since I have become very experienced in these kinds of travel, I had taken along tea, coffee, sugar, crackers, and fifteen pop-up cans of tunafish. The following day the very embarrassed local staff took some twenty of us in a bus to the nearest restaurant, that of the Academy, a couple of miles away. The restaurant was open but there was no food, and there would not be any, not for twenty people. The busload went searching for food in Kiev, after dropping us off. When they returned, one of my colleagues from Moscow State University brought us a bottle of milk (which was only slightly "off"), some bread, and some cheese.

Then the school began, and it was totally in English. Even the question-and-answer periods had to be conducted in English. This was possibly for the reason given: the organizers were anxious to have the younger scientists learn to survive in international conferences outside the USSR, which are always held in English. It might also have been that the local organizers, reputed to be strong Ukrainian nationalists, were trying to have as little Russian spoken as possible. The result was quite ghastly. The younger people typically spoke very little English, and they painfully read aloud the fifty lines in small script that they had written on transparencies. Wally gave a talk on his work, and I gave a couple of long lectures on mine. After three days, Wally went to Moscow to sightsee and to meet with two scientists in his field. He then flew home.

I stayed and went native. Though my accent remained miserable, I began to think in Russian and, amazingly, to forget English words. It was my most interesting trip to the USSR. People were no longer afraid to talk, and talk they did. They were almost unanimously angry at the Communist party and what it had done to them for seventy years. Only two of my friends, *pur et dur* Communists, deplored the changes. No one thought much of Gorbachev. "He talks a lot, that one. To have a vacation he goes and talks to foreign leaders." The staff people were angry at the increase in black marketeering, and at inflation. They were having trouble finding food, and they expected things to get worse. They did not see how they could survive. And they were incensed at the way the government had reacted to Chernobyl. They still did not have adequate access to information about its effects. They told me that when the wind shifted, Kiev had been covered by white dust, and that the population was treated cavalierly. One of the principal scientists in charge of cleaning up Chernobyl told me that even he did not yet have full access to information that he needed. I said that it had been published in the West, and described it. Yes, the government had disclosed matters to the West that it was still unwilling to tell its own people. At his request I sent him several reprints when I returned home.

I spent a lot of time in Kiev speaking with my friends Anna and Vadim Volkov. I had first met Vadim in Dubna in 1966, but I began to know him well at an international conference in Munich in 1973. He was presenting an invited paper, and he asked me to help with its English translation, which was not colloquially correct. We sat on a park bench for a couple of hours, intensely discussing the translation and the reasons for the corrections I was making. I noticed, with amusement, several hard-eyed characters who occasionally came to eavesdrop on our conversation. I finally decided that one of them was "his," one "mine," and one "German." The Cold War was still on. Gradually we came to be friends, although it took some time to clear some misconceptions: at a meeting in Canada he said that he had been told that I had signed a petition to free some prisoners-of-conscience in the USSR and he asked me whether my government had ordered me to do it. I laughed and said, "Vadim, can you imagine *anyone* telling me what to do?" By that time he knew me well enough to smile. We have remained friends, Anna and Vadim, and Wally and I, meeting at conferences and at Dubna.

The school participants made a couple of excursions to Kiev and to a park nearby. During our visit to a museum, the building began to move. It turned out that a strong earthquake, centered in Romania, had shaken the western part of the USSR all the way to Leningrad. There was no damage in Kiev but gossip reported many deaths in Romania. One of the Romanian participants was, of course, extremely upset. She tried to call home, but the connection did not go through. She couldn't find the number of the Romanian Embassy in Moscow because telephone books still did not exist. I finally called the U.S. Embassy, whose number I memorize whenever I travel to Russia, and I explained to the duty officer that she might find my story a bit weird. I said that, despite my accent, I am an American scientist, that I am at Kiev, that one of my colleagues is from Romania, that because of the earthquake she needs the number of the Romanian Embassy, and could she please give it to me. "Of course," she said, and my colleague was able to find out that her town had no casualties.

Finally I was due to leave. The organizers had insisted on paying for my air ticket from Kiev to Moscow. A couple of times I inquired whether the ticket was okay, and they told me sure, but I was concerned because the officials looked uncomfortable. When the day came, I was told that they would put me on the overnight train. I asked why, and they kept insisting that the train, really, was much nicer. I found out later that Aeroflot had instituted a regulation that tickets for foreigners, even when they are guests of local groups, had to be paid for in hard currency. When they mentioned the train, I looked at them rather coldly, and said that I would like a compartment to myself. No problem, they said, and, even if someone was assigned to share my compartment, they were sure that it would be a woman. It was, of course, a man, and a particularly suspect one at that. He claimed that he was a Syrian medical student but his story for being on the train did not hang together. It was hot, the toilets stank, the road bed was incredible, and I did not sleep. But in the corridor I met several Americans, members of a People-to-People group. They asked me to translate what they were saying to a very nice Russian. They were trying hard to tell him everything that was wrong with our country, with vituperative asides from me, but when they began to bemoan our less than universal health service, they lost his attention completely. The Russians realize fully that their medical system is awful, and they tend to idealize the West.

In Moscow, the niece of one of my friends met me and took me to the hotel of the Academy of Sciences, where the conference organizers had arranged for me to stay one night. It is surely the worst place I have ever stayed, anywhere. My cubicle had a narrow dirty bed, covered by torn sheets; the thin walls were covered with torn paper; and the facilities I shared with two men were primitive and dirty. The window gave onto the Lenin Prospekt, which has a great deal of loud traffic, but at least I always carry ear plugs. It was also hot. During the night, I tilted over the narrow bed while asleep, found myself on the dirty floor, and burst out laughing. The hotel dining room was supposed to be open for lunch. However, its door was locked and while I could hear people inside, my knocking had no effect. Then a surly-looking Russian came and started pounding on the door. It was finally opened and he demanded access. I followed him and was seated at his table. Eventually we were awarded some food.

I took a cab for a couple of hours and revisited Moscow. I bought a fresh supply of Vysotsky records. On earlier trips they had been virtually unobtainable: I had bought most of them in Paris and London. Adults did not smile in Moscow and in Kiev; only children did. The anger of the people I saw and talked with was palpable. I made arrangements for a cab to take me to the airport the next day. It was well that I did so. When I reconfirmed, the concierge said that there were no cabs and there would not be any. I explained that her colleague had made arrangements the previous day. She sourly called the cab dispatcher, and said, okay, there would be a cab. And there was one, at the correct time. I took along a Turkish scientist and his Russian coworker because they had been stranded.

At Sheremetyevo, the chaos was complete. No one was exerting any control over the hundreds of travelers, most of whom had a lot of luggage and many friends. I found a trolley and discovered which gate I had to go through to complete customs and passport examinations. It was two hours before my plane would leave, but I took a place in the queue and started shoving and yelling at people who tried to bypass us. Several of those who succeeded were clearly from the southern regions of the Soviet Union. They were the butt of very unpleasant racial comments by the Russians in the queue. Near the departure gate I watched several hundred Japanese spend their transit time in the duty-free shops, a kitten gamboling near the departure gates, and scores of birds flying around the aircraft engines. The trip to Paris on Air France was great. My first glance at the Herald Tribune was less essential than usual when leaving the USSR, because television, on the whole, had been informative. In Paris, I was picked up and taken to my favorite hotel, where I sighed happily, called Wally, took a couple of baths and went swimming. I recovered with a few days of R and R in Paris (but NOT *aux frais de la Princesse*).

5. Penn Is Evolving

Under the prodding of the Affirmative Action policies of the federal and the state governments, the University of Pennsylvania, and indeed all other universities, have changed greatly over the past twenty years. There are many women faculty throughout the university, although the Committee on the Faculty (of which I am a member) has found that in some sectors of the university, for instance in the natural sciences and in mathematics, women are underrepresented compared to the pool of women holding Ph.D.s in those fields. It is extremely difficult to solve this problem since the process by which departments search for new faculty (and promote them to tenure) is, even at the best of times, byzantine. What Affirmative Action guidelines require is that faculty positions be widely *advertised* (not simply through the old-boys' network, in which a professor calls one of his pals and says "Say, we have a faculty position. Do you know of any good men for it?" [even I used to receive such calls regularly!]) and that women (and underrepresented minority) applicants be considered by the same standards as are white, male applicants. Then if, and only if, the female (or minority) applicant is at least as "good" as the male candidate, then the female (or minority) applicant should be hired. (In nursing, where women are in the majority on the faculty, males are to be considered under Affirmative Action guidelines.)

The advertisement issue is codified and simple. The "consideration" is far more subtle; and it can easily, and sometimes unconsciously, be subverted. Many of us could write manuals on how it is done. I will just give one example. An appointment was going to be made at a very high level within one of the science departments. Several very eminent women were suggested to the Search Committee. I was assured by one of the senior members of the committee that they had been contacted and that none had been interested in the job. I met one of the women several months later and I asked her whether she had been queried as to whether she might be interested in coming to Penn. She said, no, she had not been.

And then she paused and said that it was conceivable that in a totally informal way, at some meeting or other, it could have been mentioned; but this had not been done in a serious and appropriate fashion. (She also told me that she was really not willing to come to Penn.) Were the other approaches made as casually? This was an obviously egregious case. Those involving junior faculty are far harder to unearth. Since individual departments are too often unwilling to demonstrate a real commitment to Affirmative Action, its implementation requires a dean and a provost who believe in it, and an Affirmative Action officer who scrutinizes departmental faculty decisions with savvy and toughness.

Still, life has changed at Penn, and at other universities, although the changes have been achingly slow. The English department, which until the middle seventies had no tenured women on its staff despite the large number of excellent women in the field, is justly proud of the many distinguished women faculty members it now enjoys. In the early seventies, the department had failed to promote a woman to tenure. Some of the papers in the Phyllis Rackin case were published: they included a letter from the English faculty to the dean saying that, in denying tenure, they were behind their chairman "to a man."

Yes, there have been changes. Our new dean is a woman, as are a few chairs. Several women have served as chairs of the Faculty Senate. And in 1990, the Senate passed a resolution that it was in favor of Affirmative Action; this only twenty years after it became the law of our land. The way it happened is revealing. Several of us, from throughout the university, wrote a public letter urging the Senate to do this. I took copies of the letter to half a dozen of my most enlightened male colleagues in the physics department, and gently asked them if they would consider signing it, and I left it with them. With the exception of Wally, none did. I am sure that each had a perfectly reasonable exception to the letter. A word here or there might not have been the perfect one to use. It is not that I think for a moment that these particular people are against Affirmative Action. My guess is that they didn't want their brethren to conclude that they consider Affirmative Action to be an important issue. Still, last year the department passed overwhelmingly a statement that "the Department of Physics recognizes the need to attract highly competent women (and minority) faculty. The fields of Astronomy and Astrophysics represent particularly attractive opportunities, since many women are active in these two fields. The Department will discharge all its Affirmative Action responsibilities in filling these and any other future appointments." I proposed such a statement, but the tenor of the discussion was, "What's the problem?" and "It isn't necessary!" I think it would have been defeated had not one of my colleagues, Ralph Amado, said, "Look, we may not think it is necessary, but one of our highly respected senior faculty wants it. We should therefore do it."

The physics department at Penn is not more insensitive than most others. The American Physical Society is trying to resort to a system of bribes to encourage physics departments to invite women as colloquium speakers. It offers $500 to a department that invites two female colloquium speakers per year, and $1,000 to those who invite three. I have been suggesting women as colloquium speakers for some time. This past year, when told that two of the eminent scientists I had suggested would surely be too busy to come, I requested permission to invite them myself, in the name of the department. Both accepted, and both gave spectacularly good talks. And that is really quite an important issue because faculty opinions as to what (and who) is "hot" and what is not are to some considerable extent determined by what talks they attend.

For me the issue of Affirmative Action is important, as much as anything else, because I am an elitist. I want the best possible faculty in the department and at the university. I believe that the best people will not necessarily be clones of the present staff, though many faculty are more comfortable with clones. In fact, I greatly respect most of my colleagues (save on the issue of gender) and I am very fond of several of them. The physics department at Penn is going through a rebirth as it strives to define itself for the next ten years, and it is very interesting to watch this process and to be a part of it.

In the past, all departmental votes have been "open," and a faculty member voting "no" has had to explain his or her vote openly. On matters of "essence," 85 percent of the tenured faculty has had to vote "yes" for the matter to proceed. Most of our colleagues at this, and at other universities, think of this procedure as madness. I like it because too many unreasonable decisions can be taken under the cloak of a closed written ballot. But there are, of course, some problems with an open vote. People may tend to go with the flow because to do otherwise might possibly jeopardize, at a later time, their projects or an appointment of interest to them. I don't really think that bad feelings between people are appreciably exacerbated by this "open" process: we know our colleagues' views even prior to a vote.

But it certainly was not particularly pleasant when I was the only one to vote "no" in a recent case which, in my opinion, involved Affirmative Action considerations.

However, I still feel that it is reasonable that a vote on an important issue should be explained. Incidentally, because of the Affirmative Action "problem," the university did not accept the departmental recommendation in that case. Several of my colleagues were very angry at me.

It may appear surprising that Wally and I, both individuals with very strong and sometimes contrary opinions on departmental affairs, have not allowed such differences to affect our private lives. The issue of conflict-of-interest has not arisen: our salaries and our teaching assignments have been determined by the chair of the department. Our research work is in different subfields and our research funding has been totally independent. Two years ago I resigned from a university committee which was going to consider a research proposal by Wally. But we do vote, and we are very different people. We often had opposite views on controversial decisions. Some of my colleagues may have difficulty in believing this, but we never discussed the nature of our votes either before or after a public plebiscite. We respect each other's views. In the case I mentioned earlier, in which the vote was 28 to 1, Wally recused himself from the discussion and the voting. To this day I do not know, nor do I wish to know, on which side he would have been.

Clearly the department is very important to me, and I am happy to be one of its members. Have there been any leftover problems from my own Affirmative Action fight in the early seventies? Yes, of course, there are some minor residual problems. An amusing one is that I have never been asked to give a colloquium at Penn since I became professor of physics, although I have lectured all over the world from Alma-Ata to Auckland. Despite all the evidence, I think that it is hard for some of my colleagues to really believe that my scholarly work has been significant. But I view that as their problem. It has certainly not hindered me. And I realized several years ago that my salary was relatively low. I tried, unsuccessfully, to solve the problem internally. In 1990, when the university stated that special raises would be made available to outstanding teachers and I did not receive what I considered to be an adequate raise, I appealed the departmental decision to the dean, and I did receive an appropriate salary. My low salary could have been the result of residual misconceptions, or the hard-nosed view that raises should primarily go to faculty who might otherwise leave, as I surely could not.

6. And My Life Is Changing Also

I can now concentrate on teaching physics. I have mentioned least that which is my greatest joy. I love teaching and I deeply enjoy most of my students. At Haverford College there was no question that my highest commitment should be to my students. At Penn, the reverse was more nearly true, but my commitment to teaching remained the same. To the extent that I have any philosophy of life it is that people are less likely to be barbarians if they can do well something which they enjoy, and which has some redeeming social value. I teach to help students achieve their goals. Teaching is a very emotional experience for me. I no longer get jitters before giving a scientific talk, but I am always extremely nervous giving the first lecture of the semester because I do not yet know the individuals in my "audience" and the best ways in which I can reach them.

I generally teach a course in modern physics for physics and engineering majors one semester, and an introductory course for pre-med students the other semester. The first has some fifteen very bright and interested students; the second has seventy bright and desperate students. If the latter do not receive an A or a B in the course, their chances of going to a good medical school vanish. It is a great pity that physics is used by medical schools in part to weed out applicants.

I enjoy virtually all students, whatever their motivation. Each has a unique personality, and a unique set of abilities, and eventually they leave and I never have time to get bored with them. I want to help them realize that physics is a beautiful discipline and that a scientific way of looking at the world is necessary, though, of course, not sufficient. I also would like to show them that physics is fun. At the very least I want them to learn enough physics so that they can pass the course without too much trauma, and get on with their lives.

With them I feel relaxed, after the first week. I ask them to call me by my first name, and give them my home telephone number, and I tell them that they can call me at any time if they have any questions, including on weekends. I explain about my professional work and say that occasionally I will have to go out of town, but that it will very seldom be around the times when exams are given. I tell them that I will be available to them as much as they wish. Particularly around exam times, my office and the corridor outside it are filled with students, and my colleagues look bemused and possibly think that, as a woman, I want to nurture people. But that is not it. I enjoy physics and I love to seduce my students into liking it also. I probably do not succeed with most of my pre-meds, but I think I help the physics majors realize how privileged they are to be entering the field. I feel that I enter into an informal social contract with my students: if they work very hard, if they attend classes regularly, and if they seek my help outside of class, then I must succeed in teaching them enough to do adequately in the course. I feel that it is my failure, given the quality of the students, if they do not. And, in fact, very few of the students fail.

I believe that students, while at the university, should consider their academic work to be their primary commitment. I am very unsympathetic to repeated trips out of town to athletic meets: they interfere with academic pursuits. The already generous vacation breaks are too often supplemented by additional personal trips. I am hard-nosed about enforcing those few academic regulations which have survived to this day. My view is that if they are foolish, they should be changed, and that if they remain on the books, the rules should be followed.

Recently, looking at students' course ratings I was dismayed to see how few are those courses which demand more than three to four hours of "homework" per week, and how the low-workload courses appear to be those in which virtually all students receive A's or B's (this is *not* a characteristic of natural science courses). While students should be encouraged in many ways to work hard and to do well, a structure which rewards minimal work with high grades is harmful to students and will not prepare them for life outside the university.

Since I became associate chair for Undergraduate Affairs three years ago, the number of physics majors has doubled. The main reason for this has been the revival of the Physics Club, initiated by two very energetic students. There is an undergraduate lounge which is pleasantly furnished and, during the academic year, there are usually several students in it talking to each other of science and other matters, or sleeping on the sofa. They arrange frequent talks by faculty members on their research, and yearly they have organized one-day symposia for themselves and for physics majors at all the colleges and universities in the area. I help them find research jobs during the academic year and during the summers, and I talk to them of graduate schools and of anything else they wish to discuss with an older friend who, in nonacademic areas, is totally unjudgmental.

At Thanksgiving, Wally and I throw a big bash at our home for those of our students who do not go home for the holiday, and by tradition we first go and feed the ducks and geese on the Haverford pond. And, at any time, if students I know well come to see me and ask some trivial questions, I have learned to wait patiently until they are moved to ask me the questions that are really on their mind. Over the years I have listened to long philosophical discourses while part of me is asking myself, "What does he (or she) really want to say?" and I have usually waited sympathetically until it is said. What does it have to do with teaching physics? Nothing. But it has to do with my view of being a teacher.

In the past two years three of the physics majors became graduate students at Caltech, two are at Harvard, and others are at Berkeley, Carnegie-Mellon, Chicago, Cornell, Duke, Illinois, Minnesota, Michigan, MIT, and Michigan State University. Several will take a year off before they go on. They, and the other students I have had for the past forty years, are my connection with the future, and I am proud of them. In 1991 I won a Lindback Foundation Award for Distinguished Teaching, which was totally unexpected and quite wonderful. Lindback Awards are seldom awarded to teachers of the (relatively) tough natural science courses.

The twenty seniors in the class of 1993 were the best group of physics majors I have ever encountered. I prepared a list of those students rank-ordered by their grade-point average (GPA) because the department awards a prize to the "best" senior majoring in physics. It is an interesting list: eight students have a GPA of 3.6 (out of 4.0) or above. Of these, four are women. Three of the eight students are U.S. citizens: all three are women. National "manpower" predictions are that white, U.S. males will be in the minority in the sciences and in engineering by the year 2000. It seems to me that enlightened self-interest should lead universities to appoint substantial numbers of women and underrepresented minority candidates to their faculties in science and in engineering. If they fail to do so, it is likely that their enrollments in these areas will shrink, as minorities and women enroll at those universities where they have an intrinsic support structure.

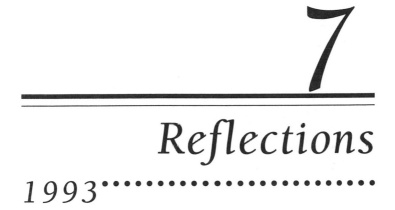

7

Reflections

1993 •

MY THOUGHTS ARE NOW PRIMARILY OF WOMEN in science, and of the character of physics—of its sociological structure as well as of its intellectual excitement. My dreams are of my friends and how best to use the time I may have left.

Reflections about Women in Science

For many years I have tried to understand the reasons why so few women are chemists, mathematicians, physicists and engineers.[14] It seems obvious to me that they are deflected from science at an early age by parental and societal factors. Then a series of what I, and others, call "microinequities" discourage many of the women who study science in college and graduate school from continuing in scientific careers. Finally, the professional lives of women scientists are far more difficult than those of male scientists, for reasons which seem to me to be remediable.

Self-confidence comes to young people from the love of their family, the respect of their peers, and from a pride in achievement. Young people, particularly young women, should be helped to understand that achieving to the best of their potential is a necessary, though surely not a sufficient, condition for feeling good about themselves. And young women should be told that popularity among their youthful peers is not as important, in the long run, as is their respect. Since pressure from conventionally reared peers in the early teens pushes young women away from science-related interests, it is very important for their families to show respect for achievement in scientific areas.[15]

Boys and young men are traditionally and repeatedly placed in competitive situations, and they learn to be judged individually, as well as in groups or teams. Fewer young women have this opportunity. Competition, both intellectual and athletic, prepares young people for real life. It is good when a young person learns repeatedly to try hard and both to lose and to win with a measure of grace, and some equanimity. One does not help young women if one shields them from competition; one makes them less able to achieve. Young women must not be shielded from success. Most women work outside the home, and it is likely that most of them would be happier doing challenging and decently paid work.

Both young women and young men should be taught to work with their hands to fix things, as well as with their minds. Understanding spatial relationships comes from manipulating objects. This is an increasingly difficult experience to acquire in a world filled with complicated black-box gadgets that cannot be opened much less understood, rather than with Model A Fords or crystal radio receivers. Shop classes should be required for all young people.

Are young women and young men treated equally in college? No, they are not. Overwhelmingly their science professors, particularly at the more prestigious universities, are males of an earlier era, an artifact of discrimination. Many of them, though consciously unaware of it, are uncomfortable with the women students in their classes. They are less likely to include women in class discussions and more likely to underestimate them. They can be intimidating and unpleasant. I believe that it is necessary that women faculty, as well as enlightened male faculty, discuss the importance of this problem with their other colleagues. It is very often the college which is at fault if a woman fails to pursue her scientific interests.

And I believe that gender discrimination is a matter of still greater importance in graduate school when relationships with older scientists, with future patrons, are first established—relationships that are critical to a scientist's entire career. I would recommend to young women that they do not accept admission to graduate departments which do not have at least a couple of women faculty members, preferably tenured, and several women graduate students. It is my view, and that of several of my women colleagues, that a woman is less likely to make it if these supports do not exist.

Science department heads must also tell male teaching assistants and male graduate students that civility and respect toward women students is expected, and that their demonstrated absence is cause for dismissal. This may be particularly important with foreign male students from male-dominated societies. In some science departments, the majority of graduate students are foreign.

I would also suggest to teachers and to employers that women need to be told explicitly that their work is good, if it is, and they need to be clearly shown respect as scientists and as individuals. Men also need that reinforcement, but they are more likely to get it as a matter of course.

In the best of all worlds, people will be treated fairly based on objective evaluations of ability and achievement. As I stated earlier, a giant step forward will occur when excellent women have no more difficulties than a man in obtaining tenured faculty positions at research universities.

And then there are the personal issues which women professionals face.

Most women scientists marry, and most marry scientists. It is usually very difficult for both to get appropriate positions in the same geographical area. The situation is improving, and some employers are making efforts to attract scientific talent and to help scientists find viable jobs for themselves and their spouses. Both members of a couple will seldom be able to achieve parity of opportunities, at a given time. Each couple has to solve this problem for itself. Based on my experience and those of many colleagues, I believe that the solutions require a great deal of mutual respect and love, and much realistic planning.

Over the years many of my women students, particularly my pre-med students, have told me that they plan to take off for a number of years to have children, and then to return to their professions. I sympathize with their desire to have children and to nurture them; and I recognize that they are realistic in not expecting their husbands to contribute much of their daily lives to their families. I do not believe, however, that it is necessary to nurture children full time. The income of a family in which both parents work full time in scientific/professional positions is high enough to permit hiring excellent individuals to take care of the day-to-day needs of children. (This is, of course, not a solution for most families, and I am very much in favor of good day-care facilities.) The bonding of children with parents can take place even more strongly if *both* parents have interesting and mind-stretching lives which they can share with them.

Because I did not have children, these remarks may appear to be unrealistic. I am basing them on my own upbringing, on the experiences of many friends, and on what I have learned of the lives of the scores of students whom I have gotten to know well.

To women who wish to become professional scientists, I am also suggesting that they remember, in a paraphrase of Hillel's words: "If you are not for yourself, who will be? And if not now, when?"

Reflections about My Colleagues and My Field

I think as well about the people I have known and the science I have seen unfold before me.

One does not "evaluate" one's personal friends but, of course, we do judge our students and our colleagues. I am amazed that, with one marginal exception, *none* of the people I didn't already respect as scientists early in my career have become recognized in their field. The reverse is not true. There have been a number of people whom I thought absolutely first-rate who disappeared from science. In some cases, personal problems and choices were at issue. Other people did not have the degree of commitment, of ruthlessness, and of single-minded determination necessary to make successful contributions to science. And some lacked luck, which is an important component of this as of all other enterprises.

I contemplate also the changes in nuclear physics. The field has now become predominantly a user-type field, where large groupings of people from different universities and government institutions work at huge central facilities whose yearly operating costs range from $10 million to over $100 million. It is no longer the individual-based field in which so many of us had such fun.

There has been a striking change also in the "geography" of the field. Before 1937 Europe dominated physics. Because of the emigration of brilliant European scientists in the thirties, and because of the Second World War, the United States became the leader for the next forty years. Now there is a certain malaise in our field, and perhaps too a lesser degree of enthusiasm and excellence. At a recent international meeting I perceived also a new arrogance on the part of European scientists toward American nuclear physics: to them it appears to be no longer at the center of things. In each of the major scientific countries—Germany, France, England, Switzerland, Belgium, Holland—there are senior scientists who more than match the best of ours, and who have received financial support from their governments which is much more effective than ours has been lately. Many of them trained in the United States and some returned home after leaving senior positions in America. The location of the research does not matter from the point of view of the advancement of science, but I am an American, and I believe that we can and should be the very best.

Final Comments

Wally has become professor emeritus. He no longer teaches but he is doing research as actively as ever. He is working on the design of an experiment to be done on the Superconducting Super Collider, the "SSC" accelerator in Texas, when it comes into operation in the year 2003. He will then be eighty-two years old. Wally still arranges MegaGlick adventures around the world for us both. We visit our friends in California, in England, in

France, and in Denmark, and they visit us. Since Marietta's death, we have become close to her daughter, Margrethe, and her granddaughter, Sophie.

Since I stopped my scholarly work, my life is more circumscribed but no less pleasant. I am greedy. I would like several more years with Wally and my students, in good health. I enjoy more than ever hearing about good science and reading excellent scientific papers. I recently saw a WNET program on Einstein, and it turned me on, particularly because of the comments in it by Abraham Pais. Pais is a supremely civilized man and an elegant physicist who has written the best of the biographies on Einstein, *Subtle Is the Lord,* and two other marvelous books. Several other people I have known were on the program: Valia Bargmann, Peter Bergmann, John Archibald Wheeler, and Eugene Wigner. In general I shy away from Nova-type programs. They are all too solemn and that isn't my view of science at all. The announcers' voices are passionless and remind me of those of the BBC's, who speak of disasters and of tea parties in the same level tones. Science is not a dead cathedral. It is live and it is *fun,* and it is full of passion. Only someone like Feynman used to be able to get this across in public. Margaret Geller, in her recent movie, *Mapping the Universe,* displays a similar mixture of charisma and science.

I now read a great deal, even more than before: the *New York Times,* a couple of books a week, and I have subscriptions to a weird and eclectic collection of American, British, and French magazines. I still teach and I continue to sit on too many professional and university panels and committees. And since Mother's death I have become very close to my sister, Iva. She does not travel and I dislike New York, so we do not see each other often but we talk several times a week on the phone, in Russian. Iva taught until she was eight-two: teaching has been as necessary for her as it has been for me.

My life has been fulfilling beyond my most unrealistic dreams. I have been stubborn, competitive, and, above all, lucky. I love Wally and I am loved by him. I am a teacher. My scientific work has been useful to my field. I have made a difference in getting women to be better accepted in my field and at my university. I have several dear friends, and there are many people of whom I am very fond.

My zest for living fully is as great as ever. I am joyous each day that I wake up next to Wally. I would like more time, but I have been privileged beyond measure. I have had a marvelous life.

Notes

1   From *La Clé des Chants,* a collection of popular songs, ed. Marie-Rose Clouzot and Pierre Jamet (Paris: Rouart Leroll and Cie, 1942).

2   While writing these remembrances I had a very strong wish to revisit the places of my childhood, and I did, in June 1991.

My husband Wally and I first went to Fontainbleau on the main Autoroute du Soleil from Paris. We approached the region of Moissy from the south. The sugar-beet fields have been replaced mostly by ZIs (*zone industrielle*), that is, by warehouses and small factories, and by malls. The center of Moissy is still dominated by the Mairie, the primary school, and the church. Our modest cottage is now surrounded by a high wall. There are heavy grilles on the first-floor windows of houses and on stores. The village has expanded into a bedroom community for the surrounding ZIs and Paris. Graffiti are plentiful. The transition from a rural to a lower-middle-class community has probably improved its plumbing but increased its tackiness. The Sucrerie de Lieusaint is derelict, with gaping holes in its roofs and windows. Only the outer walls of the workers' café remain. As for Lieusaint itself, the RN6 now jogs around its main street, and large billboards by the road describe the advantages of living in Lieusaint in new residential developments with fancy names. Our apartment house in Paris looks unchanged, but the ever-present graffiti are there too, and in the quiet streets surrounding it as well. The Lycée Victor Duruy, crumbling on the outside and finally being repaired, has become friendly and warm on the inside. There are flowers and the students look relaxed and happy. The Lycée has become co-ed. Paris itself is as beautiful as ever.

3   See Robert I. Rotberg, *Haiti: The Politics of Squalor* (Boston: Houghton-Mifflin, 1971), pp. 161–162.

4   W. A. Fowler and Fay Ajzenberg-Selove, "Thomas Lauritsen: 1915–1973," in *Biographical Memoirs*, vol. 55 (Washington, D.C.: The National Academy Press, 1985), pp. 384–396.

5   James Gleick, *Genius: The Life and Science of Richard Feynman* (New York: Pantheon, 1992).

6   *The Recollections of Eugene P. Wigner* (as told to Andrew Szanton) has recently been published (New York: Plenum, 1992). The book gives a very clear picture of this remarkable scientist. I found the description of his relationship with Gregory Breit particularly interesting.

7   V. F. Weisskopf has written a number of books; in particular, you might like to read *The Privilege of Being a Physicist* (New York: Freeman, 1989).

8   Samuel A. Goudsmit, *Alsos* (New York: Schuman, 1947).

9   Leona Marshall Libby, *The Uranium People* (New York: Crane, Russak Publishers, 1979). A book by another physicist is also of interest: it is Joan Freeman's *A Passion for Physics: The Story of a Woman Physicist* (Bristol, U.K.: Adam Hilger, 1991).

10   Luis W. Alvarez, *Adventures of a Physicist* (New York: Basic Books, 1987).

11   *Nuclear Spectroscopy. Parts A and B,* ed. F. Ajzenberg-Selove (New York: Academic Press, 1960; reprinted 1966).

12   In 1973 the chairman was anonymously "awarded" the Butterfield Award "for inadvertent service to the cause of women at Penn" for having given these reasons for denial of tenure. Alexander P. Butterfield was President Nixon's aide who, during the Watergate hearings, said, "It's all on the tapes." The existence of the tapes had previously been unknown.

13   F. Ajzenberg-Selove, "Energy Levels of Light Nuclei: A = 13 – 15," *Nuclear Physics* A523 (1991): 1–196.

14   Stephen G. Brush, "Women in Science and Engineering," *American Scientist* 79 (1991): 404–419. This general article has a very useful reference list to the literature on this subject. A recent and more specific paper is "Women in Physics: Reversing the Exclusion," by Mary Fehrs and Roman Czujko in *Physics Today,* August 1992, pp. 33–40.

15   In *Time* issue of January 20, 1992, Barbara Ehrenreich writes: "Girls' academic achievement, for example, as well as apparent aptitude and self-esteem, usually takes a nose dive at puberty. Unless nature has selected for smart girls and dumb women, something is going very wrong at about the middle-school level."

Index

About the Author

Fay Ajzenberg-Selove is one of a handful of women in her generation to become a nuclear physicist. After taking her Ph.D. at the University of Wisconsin in 1952, she did post-doctoral work at Caltech and M.I.T. and taught at Boston University, Haverford College, and the University of Pennsylvania, where she has been professor of physics since 1973. She received the Lindback Award for Distinguished Teaching in 1991. Her experimental work has resulted in a long series of papers in scientific journals and review papers as well as two books: *Nuclear Spectroscopy* and *Energy Levels of Nuclei*. A nuclear particle, the faon, was named for her by her husband. Dr. Ajzenberg-Selove has long been active in encouraging women to enter and remain in physics.

Ann Hibner Koblitz, the series editor, has won the History of Science Society's prize for outstanding work on the history of women in science. She is the author of *A Convergence of Lives: Sofia Kovalevskaia—Scientist, Writer, Revolutionary*.